VOLUME 2
CHAPTERS 9–14

INTERACTIVE ORGANIZER

INTERACTIVE ALGEBRA FOUNDATIONS

Elayn Martin-Gay

University of New Orleans

Pearson

 Pearson

ISBN-13: 978-0-13-523862-2
ISBN-10: 0-13-523862-5

Table of Contents

*Answers are available to instructors in the companion MyLab Math course instructor resource section; instructors can make answers available to students at their discretion.

VOLUME 2
CHAPTERS 9–14

INTERACTIVE ORGANIZER

Section 9.1 Symbols and Sets of Numbers

Objectives
    A Define the Meaning of the Symbols $=, \neq, <, >, \leq$, and $\geq$
    B Translate Sentences into Mathematical Statements
    C Identify Integers, Rational Numbers, Irrational Numbers, and Real Numbers

Directions: Complete your Interactive Organizer by filling in the blanks and solving exercises as you complete each screen of the Interactive Assignment.
  - For **WORK WITH ME** exercises, follow along and write each step needed and shown to solve, including the final answer.
  - For **YOUR TURN** exercises, write the exercise generated for you in MyLab Math, then "show your work" by writing each step needed to solve, including the final answer.

Objective A: Define the Meaning of the Symbols $=, \neq, <, >, \leq$, and $\geq$

Watch the objective video.

A _____ is a collection of objects called _____ or _____ .

Natural Numbers
The set of natural numbers is _____ .

Whole Numbers
The set of whole numbers is _____ .

Number line

The symbol "=" means "is _____ to."

The symbol "$\neq$" means "is _____ to."

The symbol "<" means "is _____ than."

The symbol ">" means "is _____ than."

VIDEO WORK WITH ME.

Insert <, >, or =.
7   3   0   7

The symbol "≤" means "is less than or _____ ."

The symbol "≥" means "is _____ or equal to."

VIDEO WORK WITH ME.
Is $11 \leq 11$ a true or false statement?

YOUR TURN #1:   **YOUR TURN #2:**

YOUR TURN #3:

Objective B: Translate Sentences into Mathematical Statements

Watch the objective video.

VIDEO WORK WITH ME.

Five is greater than or equal to four.   Fifteen is not equal to negative two.

YOUR TURN #1:   **YOUR TURN #2:**

Objective C: Identify Integers, Rational Numbers, Irrational Numbers, and Real Numbers

Watch the objective video.

Integers
The set of **Integers** is _____ .

Rational Numbers

$$\left\{\frac{a}{b}\middle| a \text{ and } b \text{ are integers and } \underline{\hspace{2cm}}\right\}$$

Irrational Numbers

The set of irrational numbers is the set of all non-rational numbers that correspond to _____ on the number line.

Real Numbers

The set of real numbers is the set of _____ numbers that correspond to points on the number line.

VIDEO WORK WITH ME.

Tell which set(s) each number belongs to:

$$0 \qquad\qquad \frac{2}{3}$$

Natural Numbers
Whole Numbers
Integers
Rational Numbers
Irrational Numbers
Real Numbers

YOUR TURN #1:   **YOUR TURN #2:**

Section 9.2 Properties of Real Numbers

Objectives
     A Use the Commutative and Associative Properties
     B Use the Distributive Property
     C Use the Identity and Inverse Properties

Directions: Complete your Interactive Organizer by filling in the blanks and solving exercises as you complete each screen of the Interactive Assignment.
- For **WORK WITH ME** exercises, follow along and write each step needed and shown to solve, including the final answer.
- For **YOUR TURN** exercises, write the exercise generated for you in MyLab Math, then "show your work" by writing each step needed to solve, including the final answer.

Objective A: Use the Commutative and Associative Properties

Order does not matter when adding numbers. For example, $7 + 5$ is the same as $5 + 7$; both 12. This property is given a special name – the _____ .

Also, order does not matter when multiplying numbers. For example, $-5(6)$ is the same as $6(-5)$; both -30. This property means that multiplication is commutative also and is called the

_____ .

Commutative Properties (Order)
     Addition:
     Multiplication:

Helpful Hint
Is subtraction also commutative? Try an example.

Is $3 - 2 \overset{?}{=} 2 - 3$?
     $1 \overset{?}{=} -1$ No!

There is no commutative property of _____.

Similarly, there is no commutative property of _____ . For example, $10 \div 2 \neq 2 \div 10$.

Let's now discuss grouping numbers. The way in which addition of numbers are grouped or associated does not change their sum. For example, $2 + (3 + 4)$ is the same as $(2 + 3) + 4$; both 9. Thus, $2 + (3 + 4) = (2 + 3) + 4$. This property is called the _____ .

Also, changing the grouping of numbers when multiplying does not change their product. For example, $2 \cdot (3 \cdot 4)$ is the same as $(2 \cdot 3) \cdot 4$; both 24. This is the _____ .

Associative Properties (Order)
        Addition:
        Multiplication:

Helpful Hint
Remember:

The commutative properties have to do with the _____ of numbers.

The associative properties have to do with the _____ of numbers.

WORK WITH ME #1.
a. Use a commutative property to complete the statement. $x + 16 =$ _____
b. Use a commutative property to complete the statement. $xy =$ _____
c. Use an associative property to complete the statement. $(xy) \cdot z =$ _____
d. Use an associative property to complete the statement. $(a + b) + c =$ _____

WORK WITH ME #2.
Which of the pairs is commutative?

WORK WITH ME #3.
Use the commutative and associative properties to simplify each expression.
a. $8 + (9 + b)$
b. $4(6y)$

YOUR TURN #1:   **YOUR TURN #2:**

YOUR TURN #3:   **YOUR TURN #4:**

<u>**Objective B: Use the Distributive Property**</u>

The _____ allows us to write a product as a sum or a
sum as a product.

216

WORK WITH ME #1.

WORK WITH ME #2.
Use the distributive property to write each expression without parentheses. Then simplify the result, if possible. For part d, use the distributive property to write the sum as a product.
a. $3(6 + x)$
b. $-(r - 3 - 7p)$
c. $-9(4x + 8) + 2$
d. $11x + 11y$

YOUR TURN #1:   **YOUR TURN #2:**

YOUR TURN #3:   **YOUR TURN #4:**

Objective C: Use the Identity and Inverse Properties

Identity Properties:

The number 0 is called the _____ because when 0 is added to any real number, the result is the same real number. In other words, the *identity* of the real number is not changed.

The number 1 is called the _____ because when a real number is multiplied by 1, the result is the same real number. In other words, the *identity* of the real number is not changed.

Identities for Addition and Multiplication

0 is the identity element for addition.
      $a + 0 =$ and $0 + a =$

1 is the identity element for multiplication.
      $a \cdot 1 =$ and $1 \cdot a =$

Additive Inverses or Opposites:
Two numbers are called additive inverses or opposites if their sum is _____ .

The additive inverse or opposite of 6 is –6 because _____ .

The additive inverse or opposite of –5 is 5 because _____ .

Reciprocals or Multiplicative Inverses:
Two non-zero numbers are called reciprocals or multiplicative inverses if their product is _____ .

The reciprocal or multiplicative inverse of $\frac{2}{3}$ is $\frac{3}{2}$ because _____ .

The reciprocal of –5 is $-\frac{1}{5}$ because _____ .

Additive or Multiplicative Inverses
The numbers a and $-a$ are additive inverses or opposites of each other because their sum is 0; that is,

The numbers b and $\frac{1}{b}$ (for $b \neq 0$) are reciprocals or multiplicative inverses of each other because their product is 1; that is,

WORK WITH ME #1.
Name the property illustrated by each true statement.
a. $1 \cdot 9 = 9$
b. $6 \cdot \frac{1}{6} = 1$
c. $0 + 6 = 6$

YOUR TURN #1:   **YOUR TURN #2:**

YOUR TURN #3:

Section 9.3 Further Solving Linear Equations

Objectives
    A Apply the General Strategy for Solving a Linear Equation
    B Solve Equations Containing Fractions or Decimals
    C Recognize Identities and Equations with No Solution

Directions: Complete your Interactive Organizer by filling in the blanks and solving exercises as you complete each screen of the Interactive Assignment.
- For **WORK WITH ME** exercises, follow along and write each step needed and shown to solve, including the final answer.
- For **YOUR TURN** exercises, write the exercise generated for you in MyLab Math, then "show your work" by writing each step needed to solve, including the final answer.

Objective A: Apply the General Strategy for Solving a Linear Equation

Watch the objective video.

VIDEO WORK WITH ME.

$5(2x - 1) - 2(3x) = 1$

Solving Linear Equations in One Variable

Step 1:

Step 2:

Step 3:

Step 4:

Step 5:

Step 6:

YOUR TURN #1:

Objective B: Solve Equations Containing Fractions or Decimals

If an equation contains _____, let's clear the equation of fractions by multiplying both sides by the _____ of all denominators. By doing this, we _____ working with time-consuming fractions.

WORK WITH ME #1.

$$\frac{x}{2} - 1 = \frac{2}{3}x - 3$$

WORK WITH ME #2.

Solve the equation.
$0.50x + 0.15(70) = 35.5$

YOUR TURN #1:   **YOUR TURN #2:**

Objective C: Recognize Identities and Equations with No Solution

Not every equation in one variable has a single solution. Some equations have no solution, while others have an infinite number of solutions.

Solve. $-2x + 20 = -2x - 6$

When solving $-2x + 20 = -2x - 6$, if we add $2x$ to both sides, an equivalent equation is

$20 = $

When solving an equation in x, suppose we arrive at an equivalent equation
   with no _____ , and
   the equivalent equation is _____ , such as $20 = -6$.

There is no value for x that makes $20 = -6$ a _____ equation. Thus, we conclude that there is _____ to this equation.

Not every equation in one variable has a single solution. Some equations have no solution, while others have an infinite number of solutions.

Solve. $3(x - 4) = 3x - 12$

When solving $3(x - 4) = 3x - 12$, let's multiply to remove parentheses on the left side, then subtract $3x$ from both sides. The resulting equivalent equation is
     $-12 =$

When solving an equation in x, suppose we arrive at an equivalent equation
     with no _____ , and
     the equivalent equation is _____ , such as $-12 = -12$.

This means that one side of the equation is _____ to the other side. Thus, $3(x - 4) = 3x - 12$ is an identity and _____ real numbers are solutions.

WORK WITH ME #1.

YOUR TURN #1:   **YOUR TURN #2:**

Section 9.4 Further Problem Solving

Objectives
A Solve Problems Involving Direct Translations
B Solve Problems Involving Relationships Among Unknown Quantities
C Solve Problems Involving Consecutive Integers

Directions: Complete your Interactive Organizer by filling in the blanks and solving exercises as you complete each screen of the Interactive Assignment.

- For **WORK WITH ME** exercises, follow along and write each step needed and shown to solve, including the final answer.
- For **YOUR TURN** exercises, write the exercise generated for you in MyLab Math, then "show your work" by writing each step needed to solve, including the final answer.

Objective A: Solve Problems Involving Direct Translations

Watch the objective video.

VIDEO WORK WITH ME.

Twice the difference of a number and 8 is equal to three times the sum of the number and 3. Find the number.

YOUR TURN #1:

Objective B: Solve Problems Involving Relationships Among Unknown Quantities

General Strategy for Problem Solving
Step 1:
Step 2:
Step 3:
Step 4:

WORK WITH ME #1.
Solve.
The area of the Sahara Desert is 7 times the area of the Gobi Desert. If the sum of their areas is 4,000,000 square miles, find the area of each desert.

WORK WITH ME #2.

YOUR TURN #1:   **YOUR TURN #2:**

Objective C: Solve Problems Involving Consecutive Integers

Watch the objective video.

***VIDEO* WORK WITH ME.**

The measures of the angles of a triangle are 3 consecutive even integers. Find the measure of each angle.

YOUR TURN #1:

Section 9.5 Formulas and Problem Solving

Objectives
A Use Formulas to Solve Problems
B Solve a Formula or Equation for One of Its Variables

Directions: Complete your Interactive Organizer by filling in the blanks and solving exercises as you complete each screen of the Interactive Assignment.
- For **WORK WITH ME** exercises, follow along and write each step needed and shown to solve, including the final answer.
- For **YOUR TURN** exercises, write the exercise generated for you in MyLab Math, then "show your work" by writing each step needed to solve, including the final answer.

Objective A: Use Formulas to Solve Problems

A _____ describes a known relationship among quantities. Many formulas are given as equations.

For example, the formula
        stands for

WORK WITH ME #1.

Now let's solve for time, *t*.

To solve for *t*, we _____ both sides of the equation by _____ .

The travel time was _____ hours, or _____ hours, or _____ hours.

WORK WITH ME #2.

Solve.
  a. Convert Nome, Alaska's 14°F high temperature to Celsius.
  b. An architect designs a rectangular flower garden such that the width is exactly two-thirds of the length. If 260 feet of antique picket fencing is used to enclose the garden, find the dimensions of the garden.

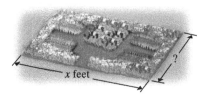

YOUR TURN #1:   **YOUR TURN #2:**

Objective B: Solve a Formula or Equation for One of Its Variables

To _____ a formula or an equation for a specified variable, we use the same steps as for solving a linear equation except that we treat the specified variable as the _____ variable in the equation.

Solving Equations for a Specified Variable

Step 1:

Step 2:

Step 3:

Step 4:

Step 5:

WORK WITH ME #1.

Solve the formula for the specified variable.
   $V = lwh$ for w

WORK WITH ME #2.

YOUR TURN #1:   **YOUR TURN #2:**

YOUR TURN #3:

Section 9.6 Linear Inequalities and Problem Solving

Objectives

      A Define Linear Inequality in One Variables, Graph Solution Sets on a Number Line, and Use Interval Notation

      B Use the Addition Property of Inequality to Solve Inequalities

      C Use the Multiplication Property of Inequality to Solve Inequalities

      D Use Both Properties to Solve Inequalities

      E Solve Problems Modeled by Inequalities

Directions: Complete your Interactive Organizer by filling in the blanks and solving exercises as you complete each screen of the Interactive Assignment.

- For **WORK WITH ME** exercises, follow along and write each step needed and shown to solve, including the final answer.
- For **YOUR TURN** exercises, write the exercise generated for you in MyLab Math, then "show your work" by writing each step needed to solve, including the final answer.

Objective A: Define Linear Inequality in One Variables, Graph Solution Sets on a Number Line, and Use Interval Notation

Watch the objective video.

Linear Inequality in One Variable

A linear inequality in one variable is an inequality that can be written in the form

where a, b, and c are _____ numbers and a is not _____.

A _____ of a linear inequality in one variable is a value for the variable that makes the inequality a _____ statement.

VIDEO WORK WITH ME.

$x \leq -1$

To write in _____ notation from a graph, follow your shading from left to _____.

_____ and _____ always have a parenthesis about them.

YOUR TURN #1:   **YOUR TURN #2:**

Objective B: Use the Addition Property of Inequality to Solve Inequalities

Watch the objective video.

Addition Property of Inequality

If a, b, and c are real numbers, then

$a < b$ and

are _____ inequalities.

VIDEO WORK WITH ME.

$x - 2 \geq -7$

YOUR TURN #1:

Objective C: Use the Multiplication Property of Inequality to Solve Inequalities

Watch the objective video.

Multiplication Property of Inequality

1. If a, b, and c are real numbers, and c is _____, then

   $a < b$ and

are equivalent inequalities.

2. If a, b, and c are real numbers, and c is _____, then

   $a < b$ and

are equivalent inequalities.

If you multiply or divide both sides of an inequality by a _____ number, you must _____ the direction of the inequality symbol.

VIDEO WORK WITH ME.

$-8x \leq 16$

YOUR TURN #1:   **YOUR TURN #2:**

Objective D: Use Both Properties to Solve Inequalities

Watch the objective video.

To Solve Linear Inequalities in One Variable
Step 1:
Step 2:
Step 3:
Step 4:
Step 5:

VIDEO WORK WITH ME.

$3(x + 2) - 6 > -2(x - 3) + 14$

YOUR TURN #1:   **YOUR TURN #2:**

Objective E: Solve Problems Modeled by Inequalities

Watch the objective video.

Some Inequality Translations			
$\geq$	$\leq$	$<$	$>$

VIDEO WORK WITH ME.

Six more than twice a number is greater than negative fourteen. Find all numbers that make this statement true.

YOUR TURN #1:

Chapter 9 Review and Practice

Study Skills
Chapter Vocabulary
Getting Ready for the Test
Review Exercises
Practice Chapter Test

Study Skills

Directions: **Watch the Study Skills video.**

Chapter Vocabulary

WORK WITH ME.

Fill in each blank with one of the words or phrases listed below:

| no solution | all real numbers | linear equation in one variable | reciprocals |

| formula | reversed | equivalent inequalities | opposites |

| linear inequality in one variable | the same |

1. A(n) _____ can be written in the form $ax + b = c$.

2. Inequalities that have the same solution are called _____ .

3. An equation that describes a known relationship among quantities is called a(n) _____ .

4. A(n) _____ can be written in the form $ax + b < c$ (or $>$, $\leq$, $\geq$).

5. The solution(s) to the equation $x + 5 = x + 5$ is/are _____ .

6. The solution(s) to the equation $x + 5 = x + 4$ is/are _____ .

7. If both sides of an inequality are multiplied or divided by the same positive number, the direction of the inequality symbol is _____ .

8. If both sides of an inequality are multiplied or divided by the same negative number, the direction of the inequality symbol is _____ .

9. Two numbers whose sum is 0 are called _____ .

10. Two numbers whose product is 1 are called _____ .

Getting Ready for the Test.

- These exercises will help you avoid common errors while taking your chapter test.

General Directions: Read the exercise Write any notes or steps in this Interactive Organizer, along with your answer to the exercise. In the MyLab Math Interactive Assignment, click the **SHOW ANSWERS** button to check your answers. Correct any errors, or press the **PLAY** button for a video solution.

Multiple Choice. Exercises 1 through 21 are Multiple Choice. Choose the correct letter.

For Exercises 1 and 2, choose two correct letters for each exercise.

1. Select the two numbers that are halfway between 2 and 3 on a number line.

    A. $3\frac{1}{2}$ B. 3.5   C. $2\frac{1}{2}$ D. 2.5

2. Select the two numbers that are halfway between –2 and –3 on a number line.

    A. $-3\frac{1}{2}$ B. –3.5   C. $-2\frac{1}{2}$ D. –2.5

For Exercises 3 through 5, choose the property that allows us to write the left side of each equals sign as the equivalent right side of the equals sign. Choice are below.

    A. distributive property   B. associative property   C. commutative property

3. $5(2x-1) = 10x - 5$ 4. $9 + x = x + 9$

5. $5(4x) = (5 \cdot 4)x$

For Exercises 6 through 9, the exercise statement and the correct answer are given. Select the correct directions.

    A. Find the opposite.   B. Find the reciprocal.   C. Evaluate or simplify.

6. 5   Answer: $\frac{1}{5}$ 7. $3 + 2(-8)$ Answer: –13

8. 2^3 Answer: 8   9. –7   Answer: 7

For Exercises 10 through 13, identify each as an

    A. equation   *or an*   B. expression

10. $6x + 2 + 4x - 10$ 11. $6x + 2 = 4x - 10$

12. $-2(x-1) = 12$ 13. $-7\left(x + \frac{1}{2}\right) - 22$

14. Subtracting $100z$ from $8m$ translates to:
   A. $100z - 8m$ B. $8m - 100z$ C. $-800zm$ D. $92zm$

15. Subtracting $7x - 1$ from $9y$ translates to:
   A. $7x - 1 - 9y$ B. $9y - 7x - 1$ C. $9y - (7x - 1)$ D. $7x - 1 - (9y)$

For Exercises 16 through 19, an equation is given. Choose the correct solution.
   A. all real number   B. no solution   C. the solution is 0

16. $7x + 6 = 7x + 9$ 17. $2y - 5 = 2y - 5$

18. $11x - 13 = 10x - 13$ 19. $x + 15 = -x + 15$

20. To solve $5(3x - 2) = -(x + 20)$, we first use the distributive property and remove parentheses by multiplying. Once this is done, the equation is
   A. $15x - 2 = -x + 20$ B. $15x - 10 = -x - 20$ C. $15x - 10 = -x + 20$ D. $15x - 7 = -x - 20$

21. To solve $\dfrac{8x}{3} + 1 = \dfrac{x - 2}{10}$ we multiply through by the LCD, 30. Once this is done, the simplified equation is
   A. $80x + 1 = 3x - 6$ B. $80x + 6 = 3x - 6$ C. $8x + 1 = x - 2$ D. $80x + 30 = 3x - 6$

Review Exercises

In the **MyLab Math, Interactive Assignment, Review Exercises** section, there are algorithmically generated "Your Turn" exercises so that you can check your knowledge of some core concepts in this chapter. Insert a few sheets of paper in your Interactive Organizer to "record and show your work" along with the final answer.

Practice Chapter Test

● These exercises will help you practice for your chapter test.

General Directions: For each exercise, "show your work" by writing each step in the solution process within your Interactive Organizer, including your final answer. In the MyLab Math Interactive Assignment, click the Show Answer button to check your answer. Correct any errors, or press the **PLAY** button for a video solution.

Translate each statement into symbols.

1.   The absolute value of a negative seven is greater than five.

2.   The sum of nine and five is greater than or equal to four.

3. Given

$$\left\{-5, -1, \frac{1}{4}, 0, 1, 7, 11.6, \sqrt{7}, 3\pi\right\},$$ list the numbers in this set that also belong to the set of:

   a. Natural numbers   b. Whole numbers
   c. Integers   d. Rational numbers
   e. Irrational numbers   f. Real numbers

Identify the property illustrated by each expression.

4. $8 + (9 + 3) = (8 + 9) + 3$

5. $6 \cdot 8 = 8 \cdot 6$

6. $-6(2 + 4) = -6 \cdot 2 + (-6) \cdot 4$

7. $\frac{1}{6}(6) = 1$

8. Find the opposite of -9.

9. Find the reciprocal of $-\frac{1}{3}$.

Use the distributive property to write each expression without parentheses. Then simplify if possible.

10. $7 + 2(5y - 3)$

11. $4(x - 2) - 3(2x - 6)$

Solve each equation.

12. $4(n - 5) = -(4 - 2n)$

13. $-2(x - 3) = x + 5 - 3x$

14. $4z + 1 - z = 1 + z$

15. $\frac{2(x + 6)}{3} = x - 5$

16. $\frac{1}{2} - x + \frac{3}{2} = x - 4$

17. $-0.3(x - 4) + x = 0.5(3 - x)$

18. $-4(a + 1) - 3a = -7(2a - 3)$

Solve each application.

19. A number increased by two-thirds of the number is 35. Find the number.

20. A gallon of water seal covers 200 square feet. How many gallons are needed to paint two coats of water seal on a deck that measures 20 feet by 35 feet?

20 feet   35 feet

233

21. Find the value of x if $y = -14$, $m = -2$, and $b = -2$ in the formula $y = mx + b$.

Solve the equation for the indicated variable.

22. $V = \pi r^2 h$ for h

23. $3x - 4y = 10$ for y

Solve the inequality. Graph the solution set and write it in interval notation.

24. $3x - 5 > 7x + 3$

Solve each inequality. Write each answer using interval notation.

25. $-0.3x \geq 2.4$

26. $-5(x - 1) + 6 \leq -3(x + 4) + 1$

27. $\dfrac{2(5x + 1)}{3} > 2$

28. California has more public libraries than any other state. It has 387 more public libraries than Ohio. If the total number of public libraries for these states is 1827, find the number of public libraries in California and the number in Ohio. (*Source:* Institute of Museum and Library Services)

Section 10.1 The Rectangular Coordinate System

Objectives

      A Define the Rectangular Coordinate System and Plot Ordered Pairs of Numbers

      B Graph Paired Data to Create a Scatter Diagram

      C Determine Whether an Ordered Pair Is a Solution of an Equation in Two Variables

      D Find the Missing Coordinate of an Ordered Pair Solution, Given One Coordinate of
           the Pair

Directions: Complete your Interactive Organizer by filling in the blanks and solving exercises as you complete each screen of the Interactive Assignment.

- For **WORK WITH ME** exercises, follow along and write each step needed and shown to solve, including the final answer.
- For **YOUR TURN** exercises, write the exercise generated for you in MyLab Math, then "show your work" by writing each step needed to solve, including the final answer.

Objective A: Define the Rectangular Coordinate System and Plot Ordered Pairs of Numbers

Watch the objective video.

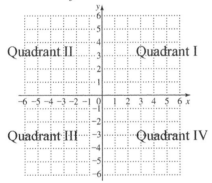

An _____ pair of numbers is of the form (x, y).

VIDEO **WORK WITH ME.**

$(1, 5), \ (-5, -2), \ (-3, 0), \ (0, -1), \ (2, -4), \ \left(-1, \ 4\frac{1}{2}\right)$

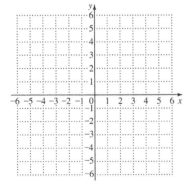

YOUR TURN #1:   **YOUR TURN #2:**

YOUR TURN #3:   **YOUR TURN #4:**

Objective B: Graph Paired Data to Create a Scatter Diagram

Watch the objective video.

VIDEO WORK WITH ME.

City	Distance from Equator (in miles)	Average Annual Snowfall (in inches)
1. Atlanta, GA	2313	2
3. Baltimore, MD	2711	21
2. Detroit, MI	2920	42
7. Miami, FL	1783	0

Write this paired data as a set of ordered pairs of the form (distance from equation, average annual snowfall).

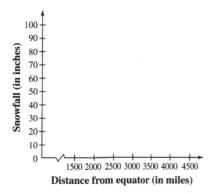

YOUR TURN #1:

Objective C: Determine Whether an Ordered Pair Is a Solution of an Equation in Two Variables

Watch the objective video.

A _____ of an equation in x and y consists of a value for x and a value for y so that a _____ statement results.

VIDEO WORK WITH ME.

$2x + y = 7$

(3, 1)   (7, 0)   (0, 7)

_____ statement: ordered pair is a solution

_____ statement: ordered pair is not a solution

YOUR TURN #1:   **YOUR TURN #2:**

Objective D: Find the Missing Coordinate of an Ordered Pair Solution, Given One Coordinate of the Pair

Watch the objective video.

VIDEO WORK WITH ME.

$x - 4y = 4$

(,–2)   (4,   )

YOUR TURN #1:

Section 10.2 Graphing Linear Equations

Objectives
A Identify Linear Equations
B Graph a Linear Equation by Finding and Plotting Ordered Pair Solutions

Directions: Complete your Interactive Organizer by filling in the blanks and solving exercises as you complete each screen of the Interactive Assignment.

- For **WORK WITH ME** exercises, follow along and write each step needed and shown to solve, including the final answer.
- For **YOUR TURN** exercises, write the exercise generated for you in MyLab Math, then "show your work" by writing each step needed to solve, including the final answer.

Objective A: Identify Linear Equations

Watch the objective video.

Linear Equations in Two Variables
A linear equation in _____ variables is an equation that can be written in the form
$Ax +$
where A, B, and C are _____ numbers and A and B are not _____ 0. The graph of a linear equation in two variables is a _____ line.

***VIDEO* WORK WITH ME.**

Determine whether each equation is a linear equation in two variables.

$x - 1.5y = -1.6$ $y = -2x$ $x + y^2 = 9$ $x = 5$

YOUR TURN #1:   **YOUR TURN #2:**

YOUR TURN #3:

Objective B: Graph a Linear Equation by Finding and Plotting Ordered Pair Solutions

From geometry, we know that a _____ line is determined by just _____ points. Graphing a linear equation in two variables, then, requires that we find just two of its _____ many solutions. Once we do so, we _____ the solution points and _____ the line _____ the points. Usually, we find a _____ solution as well, to _____ the placement of our line.

WORK WITH ME #1.

Graph the linear equation.

$x = -3y$

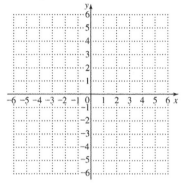

WORK WITH ME #2.

Graph the linear equation.

$-5x + 3y = 15$

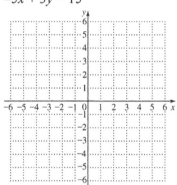

WORK WITH ME #3.

Graph the linear equation.

$y = \dfrac{1}{2}x + 2$

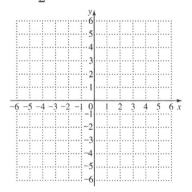

YOUR TURN #1:   **YOUR TURN #2:**

YOUR TURN #3:

Section 10.3 Intercepts

Objectives
A Identify Intercepts of a Graph
B Graph a Linear Equation by Finding and Plotting Intercepts
C Identify and Graph Vertical and Horizontal Lines

Directions: Complete your Interactive Organizer by filling in the blanks and solving exercises as you complete each screen of the Interactive Assignment.

- For **WORK WITH ME** exercises, follow along and write each step needed and shown to solve, including the final answer.
- For **YOUR TURN** exercises, write the exercise generated for you in MyLab Math, then "show your work" by writing each step needed to solve, including the final answer.

Objective A: Identify Intercepts of a Graph

Watch the objective video.

VIDEO WORK WITH ME.

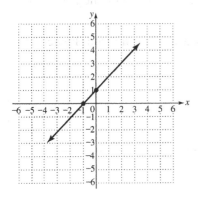

A point where a graph crosses the *y*-axis is called a _____.

A point where a graph crosses the *x*-axis is called an _____.

VIDEO WORK WITH ME.

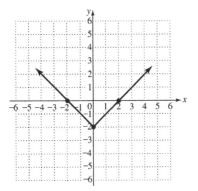

x-intercepts have _____ of 0.

y-intercepts have _____ of 0.

YOUR TURN #1:

Objective B: Graph a Linear Equation by Finding and Plotting Intercepts

When graphing a linear equation, _____ are usually easy to find since one _____ is 0.

To find the *y*-intercept of a line, given its equation, let _____, since a point on the *y*-axis has an *x*-coordinate of 0.

To find the *x*-intercept of a line, let _____, since a point on the *x*-axis has a *y*-coordinate of 0.

Finding *x*- and *y*-intercepts

To find the *x*-intercept, let _____ and solve for *x*.

To find the *y*-intercept, let _____ and solve for *y*.

WORK WITH ME #1.

Graph the linear equation by finding and plotting its intercepts.

$y = -2x$

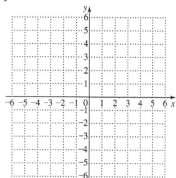

WORK WITH ME #2.

Graph the linear equation by finding and plotting its intercepts.

$4x = 3y - 9$

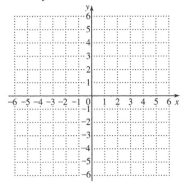

YOUR TURN #1:   **YOUR TURN #2:**

Objective C: Identify and Graph Vertical and Horizontal Lines

Watch the objective video.

VIDEO WORK WITH ME.

$y = 5$

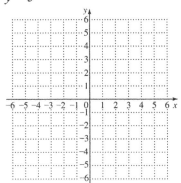

Horizontal Lines

The graph of _____, where c is a real number, is a horizontal line with y-intercept _____.

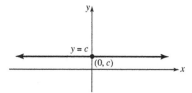

VIDEO WORK WITH ME.

$x = -3$

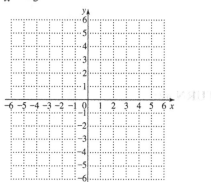

Vertical Lines

The graph of _____, where c is a real number, is a vertical line with x-intercept _____.

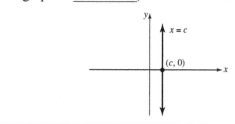

YOUR TURN #1:   **YOUR TURN #2:**

Section 10.4 Slope and Rate of Change

Objectives
   A Find the Slope of a Line Given Two Points of the Line
   B Find the Slope of a Line Given Its Equation
   C Find the Slopes of Horizontal and Vertical Lines
   D Compare the Slopes of Parallel and Perpendicular Lines
   E Interpret Slope as a Rate of Change

Directions: Complete your Interactive Organizer by filling in the blanks and solving exercises as you complete each screen of the Interactive Assignment.
   • For **WORK WITH ME** exercises, follow along and write each step needed and shown to solve, including the final answer.
   • For **YOUR TURN** exercises, write the exercise generated for you in MyLab Math, then "show your work" by writing each step needed to solve, including the final answer.

Objective A: Find the Slope of a Line Given Two Points of the Line

A key feature of a line is its _____ or _____ . In mathematics, the tilt or slant of a line is formally known as its _____ . We measure the slope of a line by the ratio of _____ change to the corresponding _____ change as we move along the line.

$$\text{slope} = \frac{\text{change in } y \text{ (vertical change)}}{\text{change in } x \text{ (horizontal change)}}$$

WORK WITH ME #1.

Find the slope of the line.

Helpful Hint

It makes no difference what two _____ of a line are chosen to find its slope. The slope of a line is the _____ everywhere on the line.

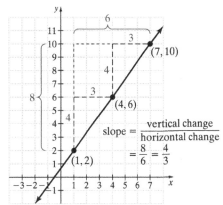

$$\text{slope} = \frac{\text{vertical change}}{\text{horizontal change}}$$
$$= \frac{8}{6} = \frac{4}{3}$$

245

In general, the _____ change or _____ between two points is the difference in the y-coordinates: $y_2 - y_1$.

The _____ change or _____ between two points is the difference of the x-coordinates: $x_2 - x_1$.

The slope of the line is the _____ of $y_2 - y_1$ to $x_2 - x_1$, and we traditionally use the letter _____ to denote slope:

$$m = \frac{y_2 - y_1}{x_2 - x_1}$$

Slope of a Line

The slope m of the line containing the points (x_1, y_1) and (x_2, y_2) is given by

$$m = \frac{\text{rise}}{\text{run}} = \frac{\text{change in } y}{\text{change in } x} = \frac{y_2 - y_1}{x_2 - x_1}, \quad \text{as long as } x_2 \neq x_1$$

WORK WITH ME #2.

Find the slope of the line.

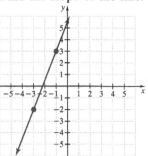

WORK WITH ME #3.

Find the slope of the line through (–1, 5) and (2,–3). Graph the line.

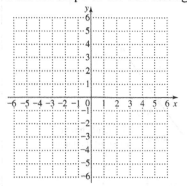

Helpful Hint

When finding slope, it makes no difference which point is identified as (x_1, y_1) and which is identified as (x_2, y_2). Just remember that whatever y-value is first in the numerator, its corresponding x-value must be first in the denominator. Another way to calculate the slope in the example above is:

$$m = \frac{y_2 - y_1}{x_2 - x_1} = \frac{5 - (-3)}{-1 - 2} = \frac{8}{-3} \text{ or } -\frac{8}{3}$$

WORK WITH ME #4.

Find the slope of the line that passes through the given points.
a. (–4, 3) and (–4, 5)
b. (5, 1) and (–2, 1)

YOUR TURN #1:   **YOUR TURN #2:**

YOUR TURN #3:

Objective B: Find the Slope of a Line Given Its Equation

Watch the objective video.

Slope-Intercept Form

When a linear equation in _____ variables is written in slope-intercept form,

$y =$

m is the slope of the line and $(0, b)$ is the y-intercept of the line.

VIDEO WORK WITH ME.

Find the slope of
$2x + y = 7$ $2x - 3y = 10$

YOUR TURN #1:   **YOUR TURN #2:**

Objective C: Find the Slopes of Horizontal and Vertical Lines

Watch the objective video.

VIDEO WORK WITH ME.

Find the slope:
$x = 1$ $y = -3$

YOUR TURN #1:   **YOUR TURN #2:**

Objective D: Compare the Slopes of Parallel and Perpendicular Lines

Watch the objective video.

Parallel Lines

Nonvertical parallel lines have the same slope and different *y*-intercepts.

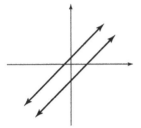

Perpendicular Lines

If the product of the slopes of two lines is −1, then the lines are perpendicular.

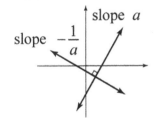

Two nonvertical lines are perpendicular if the slope of one is the negative reciprocal of the slope of the other.

VIDEO **WORK WITH ME.**

$y = \dfrac{2}{9}x + 3$

$y = -\dfrac{2}{9}x$

VIDEO WORK WITH ME.

$10 + 3x = 5y$
$5x + 3y = 1$

YOUR TURN #1:

Objective E: Interpret Slope as a Rate of Change

Watch the objective video.

VIDEO WORK WITH ME.

Find the slope and write as a rate of change.
(5000, 2100)   (20,000, 8400)

YOUR TURN #1:

Section 10.5 Equations of Lines

Objectives

A Use the Slope-Intercept Form to Graph a Linear Equation
B Use the Slope-Intercept Form to Write an Equation of a Line
C Use the Point-Slope Form to Find an Equation of a Line Given Its Slope and a Point of the Line
D Use the Point-Slope Form to Find an Equation of a Line Given Two Points of the Line
E Find Equations of Vertical and Horizontal Lines
F Use the Point-Slope Form to Solve Problems

Directions: Complete your Interactive Organizer by filling in the blanks and solving exercises as you complete each screen of the Interactive Assignment.

- For **WORK WITH ME** exercises, follow along and write each step needed and shown to solve, including the final answer.
- For **YOUR TURN** exercises, write the exercise generated for you in MyLab Math, then "show your work" by writing each step needed to solve, including the final answer.

Objective A: Use the Slope-Intercept Form to Graph a Linear Equation

Slope-Intercept Form

When a linear equation in two variables is written in _____-_____ form,

$$y = mx + b$$

$\quad\uparrow\qquad\uparrow$

$\quad$ slope $\quad$ (0, b), y-intercept

then m is the _____ of the line and (0, b) is the _____ of the line.

WORK WITH ME #1.

Use the slope-intercept form to graph the equation $y = \dfrac{3}{5}x - 2$.

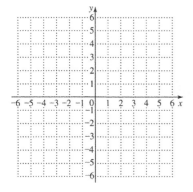

WORK WITH ME #2.

Use the slope-intercept form to graph the equation.

$$4x - 7y = -14$$

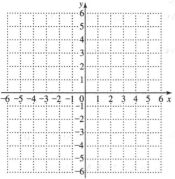

YOUR TURN #1:   **YOUR TURN #2:**

YOUR TURN #3:

Objective B: Use the Slope-Intercept Form to Write an Equation of a Line

Watch the objective video.

***VIDEO* WORK WITH ME.**

Write an equation of the line with the slope, m, and the y-intercept $(0, b)$.

$$m = -4, \quad b = -\frac{1}{6}$$

YOUR TURN #1:   **YOUR TURN #2:**

Objective C: Use the Point-Slope Form to Find an Equation of a Line Given Its Slope and a Point of the Line

Watch the objective video.

Point-Slope Form of the Equation of a Line
The point-slope form of the equation of a line is
where m is the slope of the line and (x_1, y_1) is a point on the line.

VIDEO **WORK WITH ME.**

$m = -8;\ (-1,-5)$

YOUR TURN #1:   **YOUR TURN #2:**

Objective D: Use the Point-Slope Form to Find an Equation of a Line Given Two Points of the Line

Watch the objective video.

VIDEO **WORK WITH ME.**

$(2, 3), (-1,-1)$

YOUR TURN #1:   **YOUR TURN #2:**

Objective E: Find Equations of Vertical and Horizontal Lines

Watch the objective video.

VIDEO WORK WITH ME.

Parallel to $y = 5$, through $(1, 2)$ with undefined slope, through $\left(-\dfrac{3}{4}, 1\right)$

YOUR TURN #1:   **YOUR TURN #2:**

YOUR TURN #3:

Objective F: Use the Point-Slope Form to Solve Problems

Watch the objective video.

VIDEO WORK WITH ME.

A rock is dropped from the top of a 400 foot cliff. After 1 second, the rock is traveling 32 feet per second. After 3 seconds, the rock is traveling 96 feet per second.

a. Assume that the relationship between time and speed is linear and write an equation describing this relationship. Use ordered pairs of the form (time, speed.)

b. Use this equation to determine the speed of the rock 4 seconds after it was dropped.

YOUR TURN #1:

Section 10.6 Functions

> **Objectives**
>   A Identify Relations, Domains, and Ranges
>   B Identify Functions
>   C Use the Vertical Line Test
>   D Use Function Notation

Directions: Complete your Interactive Organizer by filling in the blanks and solving exercises as you complete each screen of the Interactive Assignment.
- For **WORK WITH ME** exercises, follow along and write each step needed and shown to solve, including the final answer.
- For **YOUR TURN** exercises, write the exercise generated for you in MyLab Math, then "show your work" by writing each step needed to solve, including the final answer.

Objective A: Identify Relations, Domains, and Ranges

Watch the objective video.

A set of ordered pairs is a _____ .

The set of *x*-coordinates in a relation is called the _____ of the relation.

The set of *y*-coordinates in a relation is called the _____ of the relation.

VIDEO WORK WITH ME.

Find the domain and range of $\{(0,-2), (1,-2), (5,-2)\}$

YOUR TURN #1:

Objective B: Identify Functions

Watch the objective video.

> **Function**
>
> A _____ is a set of ordered pairs that assigns to each *x*-value exactly one *y*-value.

VIDEO WORK WITH ME.

$\{(1, 1), (2, 2), (-3,-3), (0, 0)\}$ $\{(-1, 0), (-1, 6), (-1, 8)\}$

A _____ assigns to each *x*-value exactly _____ *y*-value.

VIDEO WORK WITH ME.

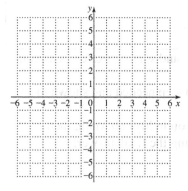 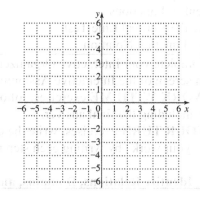

YOUR TURN #1:   **YOUR TURN #2:**

Objective C: Use the Vertical Line Test

Watch the objective video.

Vertical Line Test

If a _____ line can be drawn so that it intersects a graph _____ than once, the graph is not the graph of a function.

VIDEO WORK WITH ME.

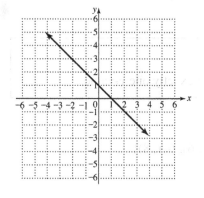 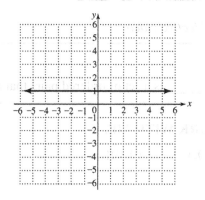

VIDEO **WORK WITH ME.**

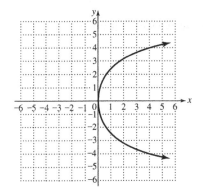 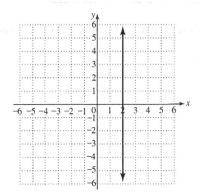

Vertical lines are _____ graphs of functions.

Recall that the graph of a linear equation is a line, and a line that is not vertical will pass the vertical line test. Thus, all linear equations are functions except those of the form $x = c$, which are vertical lines.

Show Me

Decide whether the equation describes a function.

a. $y = x$

b. $y = 2x + 1$

c. $y = 5$

d. $x = -1$

YOUR TURN #1:

YOUR TURN #2:

Objective D: Use Function Notation

The graph of the linear equation $y = 2x + 1$ passes the vertical line test, so we say that $y = 2x + 1$ is a function.

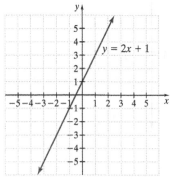

Independent and Dependent Variables

The variable y is a _____ of the variable x. For each value of x, there is only one value of y.
Thus, variable x is the _____ variable because any value in the domain can be assigned to
x. The variable y is the _____ variable because its value depends on x.

Function Notation

We can use letters such as f, g, and h to _____ functions. For example, $f(x)$ means *function
of x* and is read "___ ___ ___ " This notation is called _____ notation.

Write equation using function notation

The equation $y = 2x + 1$ can be written as _____ using function notation, and these
equations mean the same thing. In other words, _____ .

Helpful Hint

Note that $f(x)$ is a special symbol in mathematics used to denote a _____.

The _____ $f(x)$ is read "f of x." It does **not** mean $f \cdot x$ (f times x).

WORK WITH ME #1.

If $f(x) = 2x + 1$, let's find

a. $f(2)$ b. $f(-4)$

WORK WITH ME #2.

Find the domain and the range of each relation graphed.

a.

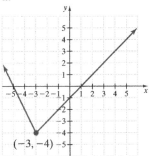

b.

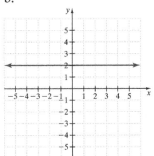

c.

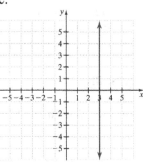

YOUR TURN #1:   **YOUR TURN #2:**

YOUR TURN #3:   **YOUR TURN #4:**

YOUR TURN #5:

Chapter 10 Review and Practice

> Study Skills
> Chapter Vocabulary
> Getting Ready for the Test
> Review Exercises
> Practice Chapter Test

Study Skills

Directions: **Watch the Study Skills video.**

Chapter Vocabulary

WORK WITH ME.

Fill in each blank with one of the words or phrases listed below:

relation	function	domain	range	standard	slope-intercept	x-intercept	x

solution	linear	slope	point-slope	y-intercept	y

1. An ordered pair is a(n) _____ of an equation in two variables if replacing the variables by the coordinates of the ordered pair results in a true statement.

2. A(n) _____ equation can be written in the form $Ax + By = C$.

3. A(n) _____ is a point of the graph where the graph crosses the *x*-axis.

4. The form $Ax + By = C$ is called _____ form.

5. A(n) _____ is a point of the graph where the graph crosses the *y*-axis.

6. The equation $y = 7x - 5$ is written in _____ form.

7. The equation $y + 1 = 7(x - 2)$ is written in _____ form.

8. To find an *x*-intercept of a graph, let _____ = 0.

9. To find an *y*-intercept of a graph, let _____ = 0.

10. The _____ of a line measures the steepness or tilt of a line.

11. A set of ordered pairs that assigns to each *x*-value exactly one *y*-value is called a _____ .

12. The set of all *x*-coordinates of a relation is called the _____ of the relation.

13. The set of all *y*-coordinates of a relation is called the _____ of the relation.

14. A set of ordered pairs is called a(n) _____ .

Getting Ready for the Test.

- These exercises will help you avoid common errors while taking your chapter test.

General Directions: Read the exercise Write any notes or steps in this Interactive Organizer, along with your answer to the exercise. In the MyLab Math Interactive Assignment, click the **SHOW ANSWERS** button to check your answers. Correct any errors, or press the **PLAY** button for a video solution.

Multiple Choice. *For Exercises 1 and 2, choose the ordered pair that is NOT a solution of the linear equation.*

1. $x - y = 5$
   A. $(7, 2)$ B. $(0, -5)$ C. $(-2, 3)$ D. $(-2, -7)$

2. $y = 4$
   A. $(4, 0)$ B. $(0, 4)$ C. $(2, 4)$ D. $(100, 4)$

3. What is the most and then the fewest number of intercepts a line may have?
   A. most: 2; fewest: 1   B. most: infinite number; fewest: 1   C. most 2: fewest: 0
   D. most: infinite number; fewest: 0

4. Choose the linear equation:
   A. $\sqrt{x} - 3y = 7$ B. $2x = 6^2$ C. $4x^3 + 6y^3 = 5^3$ D. $y = |x|$

Matching. *Match each graph in the rectangular system with its slope to the right. Each slope may be used only once.*

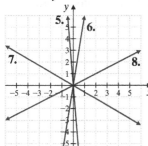

5.   A. $m = 5$

6.   B. $m = -10$

7.   C. $m = \dfrac{1}{2}$

8.   D. $m = -\dfrac{4}{7}$

Multiple Choice. *For Exercises 9 and 10, choose the best answer.*

9. An ordered pair solution for the function $f(x)$ is $(0, 5)$. This solution using function notation is:
   A. $f(5) = 0$ B. $f(5) = f(0)$ C. $f(0) = 5$ D. $0 = 5$

10. Given: $(2, 3)$ and $(0, 9)$. Final answer: $y = -3x + 9$. Select the correct instructions:
    A. Find the slope of the line through the two points.
    B. Find an equation of the line through the two points. Write the equation in standard form.
    C. Find an equation of the line through the two points. Write the equation in slope-intercept form.

Multiple Choice. *For Exercises 11 through 14, use the graph to fill in each blank using the choices below.*

      A. –2   B. 2   C. 4   D. 0   E. 3

11. $f(0) =$ _____ .

12. $f(4) =$ _____ .

13. If $f(x) = 0$, then $x =$ _____ or $x =$ _____ .

14. $f(1) =$ _____ .

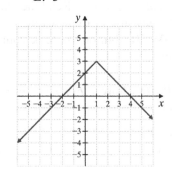

Review Exercises

In the **MyLab Math, Interactive Assignment, Review Exercises** section, there are algorithmically generated "Your Turn" exercises so that you can check your knowledge of some core concepts in this chapter. Insert a few sheets of paper in your Interactive Organizer to "record and show your work" along with the final answer.

Practice Chapter Test

- These exercises will help you practice for your chapter test.

General Directions: For each exercise, "show your work" by writing each step in the solution process within your Interactive Organizer, including your final answer. In the MyLab Math Interactive Assignment, click the Show Answer button to check your answer. Correct any errors, or press the **PLAY** button for a video solution.

Graph the following.

1. $y = \dfrac{1}{2}x$ 2. $2x + y = 8$ 3. $5x - 7y = 10$

4. $y = -1$ 5. $x - 3 = 0$

For Exercises 6 through 10, find the slope of each line.

6.

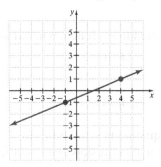

7.

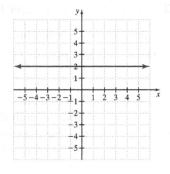

8. Through (6,–5) and (–1, 2)

9. $-3x + y = 5$

10. $x = 6$

11. Determine the slope and the *y*-intercept of the graph of $7x - 3y = 2$.

12. Determine whether the graphs of $y = 2x - 6$ and $-4x = 2y$ are parallel lines, perpendicular lines, or neither.

Find equations of the following lines. Write the equation in standard form.

13. With slope of $-\dfrac{1}{4}$, through (2, 2)

14. Through the origin and (6,–7)

15. Through (2,–5) and (1, 3)

16. Through (–5,–1) and parallel to $x = 7$

17. With slope $\dfrac{1}{8}$ and *y*-intercept (0, 12)

Determine whether each graph is the graph of a function.

18.
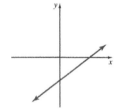

19.
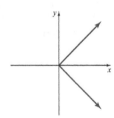

Given the following function, find the indicated function values.

20. $h(x) = x^3 - x$
   a. $h(-1)$ b. $h(0)$ c. $h(4)$

21. Find the domain of $y = \dfrac{1}{x+1}$.

*For Exercises 22 and 23, **a.** Identify the x- and y-intercepts **b.** Find the domain and the range of each function graphed.*

22.

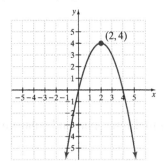

23.

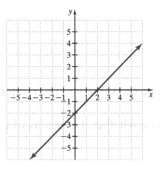

24. If $f(7) = 20$, write the corresponding ordered pair.

25. The table gives the percent of total U.S. music revenue derived from streaming music for years shown. (*Source*: Recording Industry Association of America)

Year	Percent of Music Revenue from Streaming
2011	9
2012	15
2013	21
2014	27
2015	34

a. Write this data as a set of ordered pairs of the form (year, percent of music revenue from streaming).

b. Create a scatter diagram of the data. Be sure to label the axes properly.

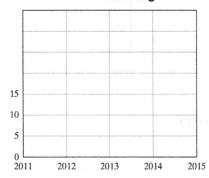

26. This graph approximates the gross box office sales *y* (in billions) for Canada and the U.S. for the year *x*. Find the slope of the line and write the slope as a rate of change. Don't forget to attach the proper units.

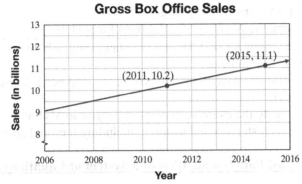

Gross Box Office Sales

Source: National Association of Theater Owners

Section 11.1 Solving Systems of Linear Equations by Graphing

Objectives
> A Determine If an Ordered Pair Is a Solution of a System of Equations in Two Variables
> B Solve a System of Linear Equations by Graphing
> C Without Graphing, Determine the Number of Solutions of a System

Directions: Complete your Interactive Organizer by filling in the blanks and solving exercises as you complete each screen of the Interactive Assignment.
- For **WORK WITH ME** exercises, follow along and write each step needed and shown to solve, including the final answer.
- For **YOUR TURN** exercises, write the exercise generated for you in MyLab Math, then "show your work" by writing each step needed to solve, including the final answer.

Objective A: Determine If an Ordered Pair Is a Solution of a System of Equations in Two Variables

Watch the objective video.

A _____ of linear equations consists of _____ or more linear equations.

A _____ of a system consists of an ordered pair that satisfies _____ equations of the system.

VIDEO **WORK WITH ME.**

$$\begin{cases} 3x - y = 5 \\ x + 2y = 11 \end{cases}$$

Is (3, 4) a solution?   Is (0, –5) a solution?

YOUR TURN #1:

Objective B: Solve a System of Linear Equations by Graphing

A solution of a system of two equations in two variables is a solution _____ to both equations. This means it is also a _____ common to the graphs of both equations. Let's find solutions of both equations in a system – that is, solutions of a system – by graphing and identifying points of intersection.

WORK WITH ME #1.

Solve the system of equations by graphing.

$$\begin{cases} 2x + y = 0 \\ 3x + y = 1 \end{cases}$$

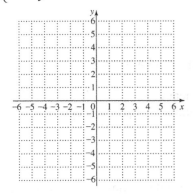

Helpful Hint

_____ drawn graphs can help when you are _____ the solution of linear equations by graphing.

Three different situations can occur when graphing the two lines associated with the two equations in a linear system:

One point of intersection:
    one solution
Consistent System
(at least one solution)
Independent Equations
(graphs of equations differ)

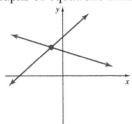

Parallel lines:
  no solution
Inconsistent System
(no solution)
Independent Equations
(graphs of equations differ)

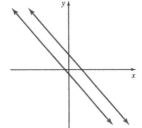

Same line:
  infinite number of solutions
Consistent System
  (at least one solution)
Dependent system
(graphs of equations identical)

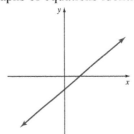

YOUR TURN #1:   **YOUR TURN #2:**

Objective C: Without Graphing, Determine the Number of Solutions of a System

Graphing alone is not an _____ way to solve a system of linear equations. We can decide _____ how many solutions a system has by writing each equation in slope-intercept form.

WORK WITH ME #1.

Without graphing, decide:
- Are the graphs of the equations identical lines, parallel lines, or lines intersecting at a single point?
- How many solutions does the system have?

a. $\begin{cases} 4x + y = 24 \\ x + 2y = 2 \end{cases}$ b. $\begin{cases} x + y = 4 \\ x + y = 3 \end{cases}$

YOUR TURN #1:   **YOUR TURN #2:**

Section 11.2 Solving Systems of Linear Equations by Substitution

Objectives
>   A Use the Substitution Method to Solve a System of Linear Equations

Directions: Complete your Interactive Organizer by filling in the blanks and solving exercises as you complete each screen of the Interactive Assignment.
- For **WORK WITH ME** exercises, follow along and write each step needed and shown to solve, including the final answer.
- For **YOUR TURN** exercises, write the exercise generated for you in MyLab Math, then "show your work" by writing each step needed to solve, including the final answer.

Objective A: Use the Substitution Method to Solve a System of Linear Equations

Graphing alone is not an _____ way to solve a system of linear equations. Now we discuss an accurate method for _____ systems of equations. This method is called the _____ method.

WORK WITH ME #1.

Solve the system of equations by the substitution method.
$$\begin{cases} x + y = 6 \\ y = -3x \end{cases}$$

Solving a System of Two Linear Equations by the Substitution Method

Step 1:

Step 2:

Step 3:

Step 4:

Step 5:

WORK WITH ME #2.

Solve the system by the substitution method.

$$\begin{cases} 7x - 3y = -14 \\ -3x + y = 6 \end{cases}$$

WORK WITH ME #3.

Solve the system by the substitution method.

a. $\begin{cases} 3x + 6y = 9 \\ 4x + 8y = 16 \end{cases}$ b. $\begin{cases} \dfrac{1}{3}x - y = 2 \\ x - 3y = 6 \end{cases}$

YOUR TURN #1:   **YOUR TURN #2:**

YOUR TURN #3:

Section 11.3 Solving Systems of Linear Equations by Addition

Objectives

    A Use the Addition Method to Solve a System of Linear Equations

Directions: Complete your Interactive Organizer by filling in the blanks and solving exercises as you complete each screen of the Interactive Assignment.

- For **WORK WITH ME** exercises, follow along and write each step needed and shown to solve, including the final answer.
- For **YOUR TURN** exercises, write the exercise generated for you in MyLab Math, then "show your work" by writing each step needed to solve, including the final answer.

Objective A: Use the Addition Method to Solve a System of Linear Equations

Substitution is an accurate way to solve a _____ system. Another accurate method for solving a system of equation is the _____ or _____ method.

The addition method is based on the addition property of _____ :

   adding _____ quantities to _____ sides of an equation does not change the _____ of the equation.

      In symbols,

           If $A = B$ and $C = D$, then $A + C = B + D$.

WORK WITH ME #1.

Solve the following system of equations.

$$\begin{cases} x + y = 7 \\ x - y = 5 \end{cases}$$

WORK WITH ME #2.

Solve the system of equations by the addition method.

$$\begin{cases} 3x - 2y = 7 \\ 5x + 4y = 8 \end{cases}$$

Solving a System of Two Linear Equations by the Addition Method

Step 1:

Step 2:

Step 3:

Step 4:

Step 5:

Step 6:

WORK WITH ME #3.

Solve the system of equations by the addition method. Clear each equation of fractions or decimals first.

$$\begin{cases} \dfrac{x}{3} - y = 2 \\ -\dfrac{x}{2} + \dfrac{3y}{2} = -3 \end{cases}$$

YOUR TURN #1:   **YOUR TURN #2:**

YOUR TURN #3:

Section 11.4 Solving Systems of Linear Equations in Three Variables

Objectives
A Solve a System of Three Linear Equations in Three Variables

Directions: Complete your Interactive Organizer by filling in the blanks and solving exercises as you complete each screen of the Interactive Assignment.

- For **WORK WITH ME** exercises, follow along and write each step needed and shown to solve, including the final answer.
- For **YOUR TURN** exercises, write the exercise generated for you in MyLab Math, then "show your work" by writing each step needed to solve, including the final answer.

Objective A: Solve a System of Three Linear Equations in Three Variables

Watch the objective video.

Solutions of this linear equation consist of ordered _____ of numbers.

Solving a System of Three Linear Equations by the Elimination Method

Step 1:

Step 2:

Step 3:

Step 4:

Step 5:

Step 6:

***VIDEO* WORK WITH ME.**

$$\begin{cases} 2x - 3y + z = 2 \\ x - 5y + 5z = 3 \\ 3x + y - 3z = 5 \end{cases}$$

YOUR TURN #1:

Section 11.5 Systems of Linear Equations and Problem Solving

Objectives
        A Solve Problems That Can Be Modeled by a System of Two Linear Equations
        B Solve Problems with Cost and Revenue Functions
        C Solve Problems That Can Be Modeled by a System of Three Linear Equations

Directions: Complete your Interactive Organizer by filling in the blanks and solving exercises as you complete each screen of the Interactive Assignment.
- For **WORK WITH ME** exercises, follow along and write each step needed and shown to solve, including the final answer.
- For **YOUR TURN** exercises, write the exercise generated for you in MyLab Math, then "show your work" by writing each step needed to solve, including the final answer.

Objective A: Solve Problems That Can Be Modeled by a System of Two Linear Equations

Problem-Solving Steps

Step 1:

Step 2:

Step 3:

Step 4:

WORK WITH ME #1.

Solve.

a. Two numbers total 83 and have a difference of 17. Find the two numbers.

b. Ann Marie Jones has been pricing Amtrak train fares for a group trip to New York. Three adults and four children must pay $159. Two adults and three children must pay $112. Find the price of an adult's ticket and find the price of a child's ticket.

WORK WITH ME #2.

The length of a rectangular road sign is 2 feet less than three times the width. Find the dimensions if the perimeter is 28 feet.

YOUR TURN #1:   **YOUR TURN #2:**

YOUR TURN #3:

Objective B: Solve Problems with Cost and Revenue Functions

Watch the objective video.

***VIDEO* WORK WITH ME.**

The planning department of Abstract Office Supplies has been asked to determine whether the company should introduce a new computer desk next year. The department estimates that $6000 of new manufacturing equipment will need to be purchased and that the cost of constructing each desk will be $200. The department also estimates that the revenue from each desk will be $450.

a. Determine the revenue function.

b. Determine the cost function.

c. Find the break-even point.

YOUR TURN #1:

Objective C: Solve Problems That Can Be Modeled by a System of Three Linear Equations

Watch the objective video.

VIDEO WORK WITH ME.

The measure of the largest angle of a triangle is 80° more than the measure of the smallest angle, and the measure of the remaining angle is 10° more than the measure of the smallest angle. Find the measure of each angle.

YOUR TURN #1:

Chapter 11 Review and Practice

> Study Skills
> Chapter Vocabulary
> Getting Ready for the Test
> Review Exercises
> Practice Chapter Test

Study Skills

Directions: **Watch the Study Skills video.**

Chapter Vocabulary

WORK WITH ME.

Fill in each blank with one of the words or phrases listed below:

| systems of linear equations | solution | consistent | independent |
| dependent | inconsistent | substitution | addition |

1. In a system of linear equations in two variables, if the graphs of the equations are the same, the equations are _____ equations.

2. Two or more linear equations are called a(n) _____ .

3. A system of equations that has at least one solution is called a(n) _____ system.

4. A(n) _____ of a system of two equations in two variables is an ordered pair of numbers that is a solution of both equations in the system.

5. Two algebraic methods for solving systems of equations are _____ and
_____ .

6. A system of equations that has no solution is called a(n) _____ system.

7. In a system of linear equations in two variables, if the graphs of the equations are different, the equations are _____ equations.

Getting Ready for the Test.

- These exercises will help you avoid common errors while taking your chapter test.

General Directions: Read the exercise Write any notes or steps in this Interactive Organizer, along with your answer to the exercise. In the MyLab Math Interactive Assignment, click the **SHOW ANSWERS** button to check your answers. Correct any errors, or press the **PLAY** button for a video solution.

1. **Multiple Choice.** *The ordered pair (−1, 2) is a solution of which system?*

A. $\begin{cases} 5x - y = -7 \\ x - y = 3 \end{cases}$

B. $\begin{cases} 3x - y = -5 \\ x + y = 1 \end{cases}$

C. $\begin{cases} x = 2 \\ x + y = 1 \end{cases}$

D. $\begin{cases} y = -1 \\ x + y = -3 \end{cases}$

Matching. For Exercises 2 through 5, match each system with the graph of its equations.

2. $\begin{cases} x = 3 \\ y = -3 \end{cases}$

3. $\begin{cases} x = -3 \\ y = 3 \end{cases}$

4. $\begin{cases} x = 3 \\ y = 3 \end{cases}$

5. $\begin{cases} x = -3 \\ y = -3 \end{cases}$

A.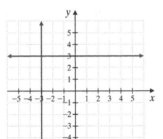

B.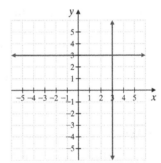

C.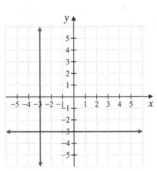

D.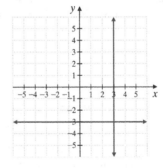

6. **Multiple Choice.** *When solving a system of two linear equations in two variables, all variables subtract out and the resulting equation is* $0 = 5$. *What does this mean?*

A. the solution is $(0, 5)$ B. the system has an infinite number of solutions

C. the system has no solution

Matching. *Match each system with its solution.   Letter choice may be used more than once or not at all.*

7. $\begin{cases} y = 5x + 2 \\ y = -5x + 2 \end{cases}$

8. $\begin{cases} y = \dfrac{1}{2}x - 3 \\ y = \dfrac{1}{2}x + 7 \end{cases}$

A. no solution
B. one solution
C. two solutions
D. an infinite number of solutions

9. $\begin{cases} y = 4x + 2 \\ 8x - 2y = -4 \end{cases}$

10. $\begin{cases} y = 6x \\ y = -\dfrac{1}{6}x \end{cases}$

Multiple Choice. *Choose the correct choice for Exercises 7 and 8. The system for these exercises is:*
$$\begin{cases} 5x - y = -8 \\ 2x + 3y = 1 \end{cases}$$

11. When solving, if we decide to multiply the first equation above by 3, the result of the first equation is
   A. $15x - 3y = -8$ B. $6x + 9y = 1$ C. $6x + 9y = 3$ D. $15x - 3y = -24$

12. When solving, if we decide to multiply the second equation above by -5, the result of the second equation is
   A. $-10x - 15y = 1$ B. $-25x + 5y = 40$ C. $-10x - 15y = -5$ D. $-25x + 5y = -8$

Review Exercises

In the **MyLab Math, Interactive Assignment, Review Exercises** section, there are algorithmically generated "Your Turn" exercises so that you can check your knowledge of some core concepts in this chapter. Insert a few sheets of paper in your Interactive Organizer to "record and show your work" along with the final answer.

Practice Chapter Test

- These exercises will help you practice for your chapter test.

General Directions: For each exercise, "show your work" by writing each step in the solution process within your Interactive Organizer, including your final answer. In the MyLab Math Interactive Assignment, click the Show Answer button to check your answer. Correct any errors, or press the **PLAY** button for a video solution.

Answer each question true or false.

1. A system of two linear equations in two variables can have exactly two solutions.

2. Although $(1, 4)$ is not a solution of $x + 2y = 6$, it can still be a solution of the system
   $$\begin{cases} x + y = 6 \\ x + y = 5 \end{cases}$$

3. If the two equations in a system of linear equations are added and the result is $3 = 0$, the system has no solution.

4. If the two equations in a system of linear equations are added and the result is $3x = 0$, the system has no solution.

Is the ordered pair a solution of the given linear system?

5. $\begin{cases} 2x - 3y = 5 \\ 6x + y = 1 \end{cases}$; $(1, -1)$

6. $\begin{cases} 4x - 3y = 24 \\ 4x + 5y = -8 \end{cases}$; $(3, -4)$

7. Use graphing to find the solutions of the system $\begin{cases} y - x = 6 \\ y + 2x = -6 \end{cases}$

8. Use the substitution method to solve the system $\begin{cases} 3x - 2y = -14 \\ x + 3y = -1 \end{cases}$

9. Use the substitution method to solve the system $\begin{cases} \dfrac{1}{2}x + 2y = -\dfrac{15}{4} \\ 4x = -y \end{cases}$

10. Use the addition method to solve the system $\begin{cases} 3x + 5y = 2 \\ 2x - 3y = 14 \end{cases}$

11. Use the addition method to solve the system $\begin{cases} 4x - 6y = 7 \\ -2x + 3y = 0 \end{cases}$

Solve each system using the substitution method or the addition method.

12. $\begin{cases} 3x + y = 7 \\ 4x + 3y = 1 \end{cases}$

13. $\begin{cases} 3(2x + y) = 4x + 20 \\ x - 2y = 3 \end{cases}$

14. $\begin{cases} \dfrac{x-3}{2} = \dfrac{2-y}{4} \\ \dfrac{7-2x}{3} = \dfrac{y}{2} \end{cases}$

Solve each problem by writing and using a system of linear equations.

15. Two numbers have a sum of 124 and a difference of 32. Find the numbers.

16. Find the amount of a 12% saline solution a lab assistant should add to 80 cc (cubic centimeters) of a 22% saline solution to have a 16% solution.

17. Texas and Missouri are the states with the most farms. Texas has 140 thousand more farms than Missouri and the total number of farms for these two states is 356 thousand. Find the number of farms for each state.

18. Two hikers start at opposite ends of the St. Tammany Trail and walk toward each other. The trail is 36 miles long and they meet in 4 hours. If one hiker walks twice as fast as the other, find both hiking speeds.

Solve.

19. $\begin{cases} 2x - 3y = 4 \\ 3y + 2z = 2 \\ x - z = -5 \end{cases}$

20. $\begin{cases} 3x - 2y - z = -1 \\ 2x - 2y = 4 \\ 2x - 2z = -12 \end{cases}$

21. The measure of the largest angle of a triangle is three less than 5 times the measure of the smallest angle. The measure of the remaining angle is 1 less than twice the measure of the smallest angle. Find the measure of each angle.

Section 12.1 Exponents

Objectives
> A Evaluate Exponential Expressions
> B Use the Product Rule for Exponents
> C Use the Power Rule for Exponents
> D Use the Power Rules for Products and Quotients
> E Use the Quotient Rule for Exponents, and Define a Number Raised to the 0 Power
> F Decide Which Rule(s) to Use to Simplify an Expression

Directions: Complete your Interactive Organizer by filling in the blanks and solving exercises as you complete each screen of the Interactive Assignment.

- For **WORK WITH ME** exercises, follow along and write each step needed and shown to solve, including the final answer.
- For **YOUR TURN** exercises, write the exercise generated for you in MyLab Math, then "show your work" by writing each step needed to solve, including the final answer.

<u>**Objective A: Evaluate Exponential Expressions**</u>

An _____ is a shorthand notation for repeated factors.

$$5^6 = \underbrace{5 \cdot 5 \cdot 5 \cdot 5 \cdot 5 \cdot 5} \qquad \text{and} \qquad (-3)^4 = \underbrace{(-3) \cdot (-3) \cdot (-3) \cdot (-3)}$$

___ factors; each factor is ___   ____ factors; each factor is _____

The _____ of an exponential expression is the repeated factor. The _____ is the number of times that the base is used as a factor.

$$5^6 \qquad\qquad (-3)^4$$

WORK WITH ME #1.

Evaluate each expression.

a. 2^3 b. 3^1 c. $(-4)^2$ d. -4^2

e. $\left(\dfrac{1}{2}\right)^4$ f. $(0.5)^3$ g. $4 \cdot 3^2$

Helpful Hint

Be careful when identifying the _____ of an exponential expression. Pay close attention to the use of _____.

WORK WITH ME #2.

Identify the base in each exponential expression.

$(-3)^2$ -3^2 $5 \cdot 3^2$

An exponent has the same meaning whether the _____ is a number or a variable.

If x is a real number and n is a positive integer,

$$x^n = \underbrace{x \cdot x \cdot x \cdot x \cdots \cdot x}$$

_____ factors of _____

WORK WITH ME #3.

Evaluate the expression given the replacement value for z.

$\dfrac{2z^4}{5}; \ z = -2$

YOUR TURN #1:   **YOUR TURN #2:**

YOUR TURN #3:   **YOUR TURN #4:**

YOUR TURN #5:   **YOUR TURN #6:**

Objective B: Use the Product Rule for Exponents

Exponential expressions can be multiplied, divided, added, subtracted, and themselves raised to powers.

WORK WITH ME #1.

$5^4 \cdot 5^3$ $x^2 \cdot x^3$

Product Rule for Exponents

If m and n are _____ integers and a is a real number, then

$a^m \cdot a^n =$

Helpful Hint
These examples will remind you of the difference between adding and multiplying terms.

Addition Examples   **Multiplication Examples**

WORK WITH ME #2.

Use the product rule to simplify.

a. $y^3 \cdot y^2 \cdot y^7$ b. $(-5)^7 \cdot (-5)^8$

c. $(5y^4)(3y)$ d. $(x^9 y)(x^{10} y^5)$

YOUR TURN #1:   **YOUR TURN #2:**

YOUR TURN #3:

285

Objective C: Use the Power Rule for Exponents

Watch the objective video.

VIDEO WORK WITH ME.

$(x^2)^3$

Power Rule for Exponents

If m and n are _____ integers and a is a real number, then

$(a^m)^n =$ _____

VIDEO WORK WITH ME.

$(x^9)^4$ $x^9 \cdot x^4$ $\left[(-5)^3\right]^7$

Helpful Hint

Take a moment to make sure that you understand when to apply the product rule and when to apply the power rule.

Product Rule--_____ Exponents	Power Rule--_____ Exponents
$x^5 \cdot x^7 =$	$(x^5)^7 =$
$y^6 \cdot y^2 =$	$(y^6)^2 =$

YOUR TURN #1:

Objective D: Use the Power Rules for Products and Quotients

Watch the objective video.

Power of a Product Rule

If n is a positive _____ and a and b are real numbers, then

$(ab)^n =$ _____

VIDEO WORK WITH ME.

$(pq)^8$

Power of a Quotient Rule

If n is a _____ integer and a and c are real numbers, then

$$\left(\frac{a}{c}\right)^n = \underline{\hspace{2cm}}, \quad c \neq 0$$

VIDEO **WORK WITH ME.**

$$\left(\frac{r}{s}\right)^9 \qquad\qquad (x^2 y^3)^5 \qquad\qquad \left(\frac{-2xz}{y^5}\right)^2$$

YOUR TURN #1:   **YOUR TURN #2:**

YOUR TURN #3:

Objective E: Use the Quotient Rule for Exponents, and Define a Number Raised to the 0 Power

Let's simplify the quotient $\dfrac{x^5}{x^3}$. Notice the numerator and the denominator have a common

_____, so we can apply the definition of an _____ and divide the numerator and
the denominator by any common _____ .

WORK WITH ME #1.

$$\frac{x^5}{x^3}$$

Quotient Rule for Exponents

If m and n are positive integers and a is a _____ number, then

$$\frac{a^m}{a^n} = \underline{\hspace{2.5cm}}$$

as long as a is not 0.

287

Let's define x^0 by simplifying $\dfrac{x^3}{x^3}$ in two ways.

Simplify Using the Quotient Rule	Simplify Using the Definition of an Exponent
$\dfrac{x^3}{x^3} =$	$\dfrac{x^3}{x^3} =$
Apply the quotient rule.	Apply the definition of an exponent.

Zero Exponent

$a^0 = 1,$ as long as a is not 0.

YOUR TURN #1:   **YOUR TURN #2:**

YOUR TURN #3:

Objective F: Decide Which Rule(s) to Use to Simplify an Expression

Watch the objective video.

VIDEO WORK WITH ME.

$(2x^3)(-8x^4)$ $\dfrac{3x^5}{x}$ $\left(\dfrac{3y^5}{6x^4}\right)^3$

YOUR TURN #1:   **YOUR TURN #2:**

YOUR TURN #3:

Section 12.2 Polynomial Functions and Adding and Subtracting Polynomials

Objectives
>   A Define Polynomial, Monomial, Binomial, Trinomial, and Degree
>   B Define Polynomial Functions
>   C Simplify a Polynomial by Combining Like Terms
>   D Add and Subtract Polynomials

Directions: Complete your Interactive Organizer by filling in the blanks and solving exercises as you complete each screen of the Interactive Assignment.
- For **WORK WITH ME** exercises, follow along and write each step needed and shown to solve, including the final answer.
- For **YOUR TURN** exercises, write the exercise generated for you in MyLab Math, then "show your work" by writing each step needed to solve, including the final answer.

Objective A: Define Polynomial, Monomial, Binomial, Trinomial, and Degree

Watch the objective video.

A _____ is a number or the product of numbers and variables raised to powers.

VIDEO **WORK WITH ME.**
$5x^2 + 6x - 2$
>   What are the terms of the expression?
>   How many terms are in the expression?

A _____ in x is a finite sum of terms of the form
> ax^n, where a is a real number and n is a whole number.

A _____ is a polynomial with exactly 1 term.

A _____ is a polynomial with exactly two terms.

A _____ is a polynomial with exactly three terms.

VIDEO **WORK WITH ME.**
What kind of polynomial is $5x^2 + 6x - 2$?

Degree of a Term
>   The degree of a term is the _____ of the exponents on the variables contained in the term.

Degree of a Polynomial
>   The degree of a polynomial is the _____ degree of any term of the polynomial.

VIDEO **WORK WITH ME.**

$-2t^2 + 3t + 6$

What kind of polynomial is $-2t^2 + 3t + 6$?

Name the terms of $-2t^2 + 3t + 6$.

What is the degree of $-2t^2 + 3t + 6$?

VIDEO **WORK WITH ME.**

$12x^4y - x^2y^2 - 12x^2y^4$

What kind of polynomial is $12x^4y - x^2y^2 - 12x^2y^4$?

Name the terms of $12x^4y - x^2y^2 - 12x^2y^4$.

What is the degree of $12x^4y - x^2y^2 - 12x^2y^4$?

YOUR TURN #1:   **YOUR TURN #2:**

YOUR TURN #3:

Objective B: Define Polynomial Functions

Watch the objective video.

VIDEO **WORK WITH ME.**

$Q(x) = 5x^2 - 1$

Find $Q(-10)$.

YOUR TURN #1:   **YOUR TURN #2:**

Objective C: Simplify a Polynomial by Combining Like Terms

Watch the objective video.

_____ terms contain the same variables raised to exactly the same powers.

Only like terms can be _____ .

VIDEO WORK WITH ME.

$14x^2 + 9x^2$ $15x^2 - 3x^2 - y$ $0.1y^2 - 1.2y^2 + 6.7 - 1.9$

YOUR TURN #1:   **YOUR TURN #2:**

YOUR TURN #3:

Objective D: Add and Subtract Polynomials

Adding Polynomials
To add polynomials, _____ all like terms.

WORK WITH ME #1.

Perform the indicated operations.

a. $(-7x + 5) + (-3x^2 + 7x + 5)$ b. $3t^2 + 4$
                                                                    $+ 5t^2 - 8$

Subtracting Polynomials
To subtract two polynomials, _____ the signs of the terms of the polynomial being subtracted and then _____ .

WORK WITH ME #2.

Perform the indicated operations.

a. $(2x^2 + 5) - (3x^2 - 9)$

b. $\begin{array}{r} 5x^3 - 4x^2 + 6x - 2 \\ -(3x^2 - 2x^2 - x - 4) \end{array}$

c. Subtract $(19x^2 + 5)$ from $(81x^2 + 10)$.

YOUR TURN #1:   **YOUR TURN #2:**

YOUR TURN #3:   **YOUR TURN #4:**

Section 12.3 Multiplying Polynomials

Objectives
> A Multiply Monomials
> B Use the Distributive Property to Multiply Polynomials
> C Multiply Polynomials Vertically

Directions: Complete your Interactive Organizer by filling in the blanks and solving exercises as you complete each screen of the Interactive Assignment.
- For **WORK WITH ME** exercises, follow along and write each step needed and shown to solve, including the final answer.
- For **YOUR TURN** exercises, write the exercise generated for you in MyLab Math, then "show your work" by writing each step needed to solve, including the final answer.

Objective A: Multiply Monomials

Watch the objective video.

Commutative Property: may _____ factors
Associative Property: may _____ factors

VIDEO WORK WITH ME.

$\left(-\dfrac{1}{3}y^2\right)\left(\dfrac{2}{5}y\right)$ Product Rule for Exponents: $a^m \cdot a^n = $ _____

YOUR TURN #1:

Objective B: Use the Distributive Property to Multiply Polynomials

To multiply polynomials that are not monomials, use the _____ property.

Multiply: $5x(2x^3 + 6) = $

WORK WITH ME #1.

Multiply.

a. $3x(2x + 5)$ b. $-y(4x^3 - 7x^2y + xy^2 + 3y^3)$

To Multiply Two Polynomials

Multiply each _____ of the first polynomial by each term of the _____ polynomial and then _____ like terms.

WORK WITH ME #2.

Multiply.

a. $(a+7)(a-2)$ b. $(3x^2+1)^2$

c. $(x+5)(x^3-3x+4)$

YOUR TURN #1:   **YOUR TURN #2:**

YOUR TURN #3:

Objective C: Multiply Polynomials Vertically

Watch the objective video.

To multiply two polynomials: Multiply _____ term of one polynomial by _____ term of the other.

VIDEO WORK WITH ME.
$(5x+1)(2x^2+4x-1)$

Like Terms: Terms with _____ variables raised to _____ powers.

YOUR TURN #1:   **YOUR TURN #2:**

Section 12.4 Special Products

Objectives

      A Multiply Two Binomials Using the FOIL Method
      B Square a Binomial
      C Multiply the Sum and Difference of Two Terms
      D Use Special Products to Multiply Binomials

Directions: Complete your Interactive Organizer by filling in the blanks and solving exercises as you complete each screen of the Interactive Assignment.

- For **WORK WITH ME** exercises, follow along and write each step needed and shown to solve, including the final answer.
- For **YOUR TURN** exercises, write the exercise generated for you in MyLab Math, then "show your work" by writing each step needed to solve, including the final answer.

Objective A: Multiply Two Binomials Using the FOIL Method

We can multiply binomials using a special order of products. Let's introduce a special order for multiplying binomials called the FOIL order or method. This FOIL method is demonstrated by multiplying $(3x + 1)$ by $(2x + 5)$.

The FOIL Method

F

O

I

L

Product

WORK WITH ME #1.

Multiply using the FOIL method.

  a. $(x + 3)(x + 4)$ b. $(3b + 7)(2b - 5)$ c. $(5x + 9)^2$

YOUR TURN #1:   **YOUR TURN #2:**

YOUR TURN #3:

Objective B: Square a Binomial

Watch the objective video.

Squaring a Binomial
A binomial squared is equal to the _____ of the first time plus or minus _____ the product of both terms plus the _____ of the second term.

$$(a+b)^2 =$$ $$(a-b)^2 =$$

VIDEO **WORK WITH ME.**
$(2x-1)^2$

YOUR TURN #1:   **YOUR TURN #2:**

Objective C: Multiply the Sum and Difference of Two Terms

Watch the objective video.

Sum and difference of same two terms: _____

VIDEO **WORK WITH ME.**
$(9x+y)(9x-y)$

Multiplying the Sum and Difference of Two Terms
The product of the sum and difference of two terms is the square of the first term minus the square of the second term.
$$(a+b)(a-b) = \underline{\hspace{3cm}}$$

297

VIDEO WORK WITH ME.

$(a-7)(a+7)$ 　　　　　　 $(4x+5)(4x-5)$ 　　　　　　 $\left(\frac{1}{3}a^2-7\right)\left(\frac{1}{3}a^2+7\right)$

YOUR TURN #1: 　　　　　　　　　　　　 **YOUR TURN #2:**

Objective D: Use Special Products to Multiply Binomials

Watch the objective video.

VIDEO WORK WITH ME.

$(3a+1)^2$ 　　　　　　　　　　　　 $(x+3)(x^2-6x+1)$

YOUR TURN #1: 　　　　　　　　　　　　 **YOUR TURN #2:**

YOUR TURN #3: 　　　　　　　　　　　　 **YOUR TURN #4:**

Section 12.5 Negative Exponents and Scientific Notation

Objectives
- A Simplify Expressions Containing Negative Exponents
- B Use All the Rules and Definitions for Exponents to Simplify Exponential Expressions
- C Write Numbers in Scientific Notation
- D Convert Numbers from Scientific Notation to Standard Form
- E Perform Operations on Numbers Written in Scientific Notation

Directions: Complete your Interactive Organizer by filling in the blanks and solving exercises as you complete each screen of the Interactive Assignment.
- For **WORK WITH ME** exercises, follow along and write each step needed and shown to solve, including the final answer.
- For **YOUR TURN** exercises, write the exercise generated for you in MyLab Math, then "show your work" by writing each step needed to solve, including the final answer.

Objective A: Simplify Expressions Containing Negative Exponents

Let's define negative exponents by simplifying $\dfrac{x^2}{x^5}$ in two ways.

Simplify Using the Quotient Rule

$$\frac{x^2}{x^5} =$$

Simplify Using the Definition of an Exponent

$$\frac{x^2}{x^5} =$$

Negative Exponent

If a is a real number other than _____ and n is an _____ , then

$$a^{-n} = \frac{1}{a^n}$$

Helpful Hint

Things to remember about negative exponents:
- A negative exponent _____ the sign of its base.
- Another way to write a^{-n} is to take its _____ and change the sign of its exponent: $a^{-n} = \dfrac{1}{a^n}$

Show Me Examples with Number Bases   **Show Me Examples with Variable Bases**

WORK WITH ME #1.

Simplify.

a. 2^{-4}

b. $\dfrac{1}{x^{-1/3}}$

c. $\dfrac{p^{-4}}{q^{-9}}$

WORK WITH ME #2.

Simplify by writing each expression with positive exponents only.

a. 3^{-2}

b. $2x^{-3}$

c. $(-2)^{-4}$

d. $2^{-1} + 4^{-1}$

e. $\dfrac{1}{7^{-2}}$

YOUR TURN #1:   **YOUR TURN #2:**

YOUR TURN #3:   **YOUR TURN #4:**

Objective B: Use All the Rules and Definitions for Exponents to Simplify Exponential Expressions

All the previously stated rules for exponents apply for negative exponents also. Here is a summary of the rules and definitions for exponents.

Summary of Exponent Rules

Below, m and n are integers, and a, b, and c are real numbers.

Product Rule

$$a^m \cdot a^n =$$

Power Rule

$$(a^m)^n =$$

Power of a Product Rule

$$(ab)^n =$$

Power of a Quotient Rule

$$\left(\frac{a}{c}\right)^n =$$

Quotient Rule

$$\frac{a^m}{a^n} =$$

Zero Exponent Rule

$$a^0 =$$

Negative Exponent Rule

$$a^{-n} =$$

WORK WITH ME #1.

Simplify the following expressions. Write each result using positive exponents only.

a. $\dfrac{(x^3)^4 \cdot x}{x^7}$

b. $\left(\dfrac{3a^2}{b}\right)^{-3}$

WORK WITH ME #2.

Simplify each expression. Write each result using positive exponents only.

a. $\dfrac{r}{r^{-3} r^{-2}}$

b. $\dfrac{\left(-2xy^{-3}\right)^{-3}}{\left(xy^{-1}\right)^{-1}}$

YOUR TURN #1:   **YOUR TURN #2:**

YOUR TURN #3:   **YOUR TURN #4:**

Objective C: Write Numbers in Scientific Notation

Watch the objective video.

To Write a Number in Scientific Notation
Step 1:
Step 2:
Step 3:

VIDEO **WORK WITH ME.**

Write each number in scientific notation.
78,000   0.00000167

YOUR TURN #1:   **YOUR TURN #2:**

Objective D: Convert Numbers from Scientific Notation to Standard Form

Watch the objective video.

Exponent on _____ tells how to write in standard form.

VIDEO WORK WITH ME.

Write each number in standard form.

3.3×10^{-2} 2.032×10^{4}

YOUR TURN #1: **YOUR TURN #2:**

Objective E: Perform Operations on Numbers Written in Scientific Notation

Watch the objective video.

VIDEO WORK WITH ME.

$$\frac{1.4 \times 10^{-2}}{7 \times 10^{-8}}$$

YOUR TURN #1: **YOUR TURN #2:**

Section 12.6 Dividing Polynomials

Objectives
A Divide a Polynomial by a Monomial
B Use Long Division to Divide a Polynomial by Another Polynomial

Directions: Complete your Interactive Organizer by filling in the blanks and solving exercises as you complete each screen of the Interactive Assignment.

- For **WORK WITH ME** exercises, follow along and write each step needed and shown to solve, including the final answer.
- For **YOUR TURN** exercises, write the exercise generated for you in MyLab Math, then "show your work" by writing each step needed to solve, including the final answer.

Objective A: Divide a Polynomial by a Monomial

Fractions that have a common denominator are added by adding the _____ :

$$\frac{a}{c} + \frac{b}{c} = \frac{a+b}{c}$$

To divide a polynomial by a monomial, read this equation from right to left with $c \neq 0$.

Dividing a Polynomial by a Monomial
Divide each term of the polynomial by the _____ .
$$\frac{a+b}{c} = \frac{a}{c} + \frac{b}{c}, \quad c \neq 0$$

WORK WITH ME #1.

Perform each division.

a. $\dfrac{12x^4 + 3x^2}{x}$

b. $\dfrac{-9x^5 + 3x^4 - 12}{3x^3}$

YOUR TURN #1:   **YOUR TURN #2:**

Objective B: Use Long Division to Divide a Polynomial by Another Polynomial

To divide a polynomial by a polynomial other than a monomial, we use _____ division. Polynomial long division is similar to number long division.

WORK WITH ME #1.

$$\frac{x^2 + 4x + 3}{x + 3}$$

WORK WITH ME #2.

a. $\dfrac{2b^3 + 9b^2 + 6b - 4}{b + 4}$ b. $\dfrac{x^3 - 27}{x - 3}$

YOUR TURN #1:   **YOUR TURN #2:**

YOUR TURN #3:

Section 12.7 Synthetic Division and the Remainder Theorem

Objectives
    A Use Synthetic Division to Divide a Polynomial by a Binomial
    B Use the Remainder Theorem to Evaluate Polynomials

Directions: Complete your Interactive Organizer by filling in the blanks and solving exercises as you complete each screen of the Interactive Assignment.
- For **WORK WITH ME** exercises, follow along and write each step needed and shown to solve, including the final answer.
- For **YOUR TURN** exercises, write the exercise generated for you in MyLab Math, then "show your work" by writing each step needed to solve, including the final answer.

Objective A: Use Synthetic Division to Divide a Polynomial by a Binomial

Watch the objective video.

When a polynomial is to be divided by a binomial of the form _____ , synthetic division, a shortcut process, may be used.

***VIDEO* WORK WITH ME.**

$(x^3 - 7x^2 - 13x + 5) \div (x - 2)$

    The last number is the _____ of the division process.

    The quotient polynomial has a degree of _____ less than the dividend polynomial.

***VIDEO* WORK WITH ME.**

$(7x^2 - 4x + 12 + 3x^3) \div (x + 1)$

Helpful Hint

Before dividing by synthetic division, check the following:

  · Make sure the dividend is written in _____ order of variable exponents.
  · Any "missing powers" of the variable should be represented by _____ times the variable raised to the missing power.

YOUR TURN #1:   **YOUR TURN #2:**

Objective B: Use the Remainder Theorem to Evaluate Polynomials

Watch the objective video.

***VIDEO* WORK WITH ME.**

$P(x) = x^2 - x + 3; \quad P(5)$

Remainder Theorem
If a polynomial $P(x)$ is divided by $x - c$, then the remainder is _____ .

***VIDEO* WORK WITH ME.**

$P(x) = 4x^4 + x^2 - 2; \quad -1$

YOUR TURN #1:   **YOUR TURN #2:**

YOUR TURN #3:

Chapter 12 Review and Practice

Study Skills
Chapter Vocabulary
Getting Ready for the Test
Review Exercises
Practice Chapter Test

Study Skills

Directions: **Watch the Study Skills video.**

Chapter Vocabulary

WORK WITH ME.

Fill in each blank with one of the words or phrases listed below:

| term | coefficient | monomial | binomial | trinomial |

| polynomial | degree of a term | distributive | FOIL | degree of a polynomial |

1. A _____ is a number or the product of numbers and variables raised to powers.

2. The _____ method may be used when multiplying two binomials.

3. A polynomial with exactly 3 terms is called a _____ .

4. The _____ is the greatest degree of any terms of the polynomial.

5. A polynomial with exactly 2 terms is called a _____ .

6. The _____ of a term is its numerical factor.

7. The _____ is the sum of the exponents on the variables in the term.

8. A polynomial with exactly 1 term is called a _____ .

9. Monomials, binomials, and trinomials are all examples of _____ .

10. The _____ property is used to multiply $2x(x-4)$.

Getting Ready for the Test.

- These exercises will help you avoid common errors while taking your chapter test.

General Directions: Read the exercise Write any notes or steps in this Interactive Organizer, along with your answer to the exercise. In the MyLab Math Interactive Assignment, click the **SHOW ANSWERS** button to check your answers. Correct any errors, or press the **PLAY** button for a video solution.

Matching. *Match the expression with the exponent operation needed to simplify. Letters may be used more than once or not at all.*

1. $x^2 \cdot x^5$ A. multiply the exponents

2. $(x^2)^5$ B. divide the exponents

3. $x^2 + x^5$ C. add the exponents

4. $\dfrac{x^5}{x^2}$ D. subtract the exponents

                                        E. this expression will not simplify

Matching. *Match the operation with the result when the operation is performed on the given terms. Letters may be used more than once or not at all.*

    Given terms: $20y$ and $4y$.

5. Add the terms   A. $80y$ E. $80y^2$

6. Subtract the terms   B. $24y^2$ F. $24y$

7. Multiply the terms   C. $16y$ G. $16y^2$

8. Divide the terms   D. 16 H. $5y$

                                          I. 5

Multiple Choice. *The expression* 5^{-1} *is equivalent to*

9. A. -5 B. 4 C. $\dfrac{1}{5}$ D. $-\dfrac{1}{5}$

Multiple Choice. *The expression* 2^{-3} *is equivalent to*

10. A. -6 B. -1 C. $-\dfrac{1}{6}$ D. $\dfrac{1}{8}$

Matching. *Match each expression with its simplified form. Letters may be used more than once or not at all.*

11. $y + y + y$ A. $3y^3$ E. $-3y^3$

12. $y \cdot y \cdot y$ B. y^3 F. $-y^3$

13. $(-y)(-y)(-y)$ C. $3y$

14. $-y - y - y$ D. $-3y$

Multiple Choice. *Choose the division exercise that can be performed using the synthetic division process.*

15.   A. $(x^3 - 5x + 15) \div (x^2 - 5)$ B. $(y^4 - y^3 + y^2 - 2) \div (y^3 + 1)$

         C. $(2x^3 - 5x^2 + 7) \div \left(x - \dfrac{1}{2}\right)$ D. $(z^5 - 4) \div (z^4 + 2z - 2)$

Review Exercises

In the **MyLab Math, Interactive Assignment, Review Exercises** section, there are algorithmically generated "Your Turn" exercises so that you can check your knowledge of some core concepts in this chapter. Insert a few sheets of paper in your Interactive Organizer to "record and show your work" along with the final answer.

Practice Chapter Test

- These exercises will help you practice for your chapter test.

General Directions: For each exercise, "show your work" by writing each step in the solution process within your Interactive Organizer, including your final answer. In the MyLab Math Interactive Assignment, click the Show Answer button to check your answer. Correct any errors, or press the **PLAY** button for a video solution.

Evaluate each expression.

1. 2^5 2. $(-3)^4$

3. -3^4 4. 4^{-3}

Simplify each exponential expression. Write the result using only positive exponents.

5. $(3x^2)(-5x^9)$ 6. $\dfrac{y^7}{y^2}$

7. $\dfrac{r^{-8}}{r^{-3}}$ 8. $\left(\dfrac{x^2 y^3}{x^3 y^{-4}}\right)^2$

9. $\dfrac{6^2 x^{-4} y^{-1}}{6^3 x^{-3} y^7}$

Express each number in scientific notation.

10. 563,000

11. 0.0000863

Write each number in standard notation.

12. 1.5×10^{-3}

13. 6.23×10^4

14. Simplify. Write the answer in standard notation.

$$(1.2 \times 10^5)(3 \times 10^{-7})$$

15.   a. Complete the table for the polynomial $4xy^2 + 7xyz + x^3y - 2$.

Term	Numerical Coefficient	Degree of Term
$4xy^2$		
$7xyz$		
x^3y		
-2		

    b. What is the degree of the polynomial?

16. Simplify by combining like terms.

$$5x^2 + 4xy - 7x^2 + 11 + 8xy$$

Perform each indicated operation.

17. $(8x^3 + 7x^2 + 4x - 7) + (8x^3 - 7x - 6)$

18. $5x^3 + x^2 + 5x - 2 - (8x^3 - 4x^2 + x - 7)$

19. Subtract $(4x + 2)$ from the sum of $(8x^2 + 7x + 5)$ and $(x^3 - 8)$.

Multiply.

20. $(3x + 7)(x^2 + 5x + 2)$

21. $3x^2(2x^2 - 3x + 7)$

22. $(x + 7)(3x - 5)$

23. $\left(3x - \dfrac{1}{5}\right)\left(3x + \dfrac{1}{5}\right)$

24. $(4x-2)^2$

25. $(8x+3)^2$

26. $(x^2-9b)(x^2+9b)$

Solve.

27. The height of the roof of the Bank of China building in Hong Kong is 1001 feet. Neglecting air resistance, the height of an object dropped from this building at time t seconds is given by the polynomial function $P(t)=-16t^2+1001$. Find the height of the object at the given times below.

t	0 seconds	1 second	3 seconds	5 seconds
$P(t)=-16t^2+1001$				

28. Find the area of the top of the table. Express the area as a product, then multiply and simplify.

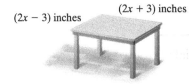

$(2x-3)$ inches $(2x+3)$ inches

Divide.

29. $\dfrac{4x^2+24xy-7x}{8xy}$

30. $(x^2+7x+10)\div(x+5)$

31. $\dfrac{27x^3-8}{3x+2}$

32. A pebble is hurled from the top of the Canada Trust Tower, which is 880 feet tall, with an initial velocity of 96 feet per second. Neglecting air resistance, the height $h(t)$ of the pebble after t seconds is given by the polynomial function

$$h(t)=-16t^2+96t+880$$

   a. Find the height of the pebble when $t=1$.
   b. Find the height of the pebble when $t=5.1$.

33. Use synthetic division to divide $(4x^4-3x^3-x-1)$ by $(x+3)$.

34. If $P(x)=4x^4+7x^2-2x-5$, use the remainder theorem to find $P(-2)$.

312

Section 13.1 The Greatest Common Factor and Factoring by Grouping

Objectives
A Find the Greatest Common Factor of a List of Integers
B Find the Greatest Common Factor of a List of Terms
C Factor Out the Greatest Common Factor from a Polynomial
D Factor a Polynomial by Grouping

Directions: Complete your Interactive Organizer by filling in the blanks and solving exercises as you complete each screen of the Interactive Assignment.
- For **WORK WITH ME** exercises, follow along and write each step needed and shown to solve, including the final answer.
- For **YOUR TURN** exercises, write the exercise generated for you in MyLab Math, then "show your work" by writing each step needed to solve, including the final answer.

<u>**Objective A: Find the Greatest Common Factor of a List of Integers**</u>

Watch the objective video.

$$2 \cdot 5 = 10$$

$$(x+4)(x+5) = x^2 + 9x + 20$$

Finding the GCF of a List of Integers
Step 1:
Step 2:
Step 3:

A _____ number is a natural number other than 1 whose only factors are 1 and itself.

VIDEO **WORK WITH ME.**

Find the GCF of
36, 90

YOUR TURN #1:

313

Objective B: Find the Greatest Common Factor of a List of Terms

Watch the objective video.

VIDEO WORK WITH ME.

Find the GCF of

$x^2, \ x^3, \ x^5$ $12y^4, \ 20y^3$

Helpful Hint

Remember that the GCF of a list of terms contains the _____ exponent on each common variable.

YOUR TURN #1:   **YOUR TURN #2:**

Objective C: Factor Out the Greatest Common Factor from a Polynomial

The first step in factoring a polynomial is to find the _____ of its terms. Once we do so, we can write the polynomial as a product by _____ out the GCF.

The polynomial _____ has two terms: _____ and _____ . The GCF of these terms is _____ . Let's factor out _____ from each term.

Thus, a factored form of $8x + 14$ is _____ .

Helpful Hint

A factored form of $8x + 14$ is *not*
$$2 \cdot 4x + 2 \cdot 7$$

The _____ have been factored (written as a product) but the _____ $8x + 14$ has not been factored (written as a product).

A factored form of $8x + 14$ is the _____ $2(4x + 7)$.

WORK WITH ME #1.

Factor out the GCF from each polynomial.

a. $14x^3y + 7x^2y - 7xy$

b. $y(x^2 + 2) + 3(x^2 + 2)$

YOUR TURN #1:   **YOUR TURN #2:**

Objective D: Factor a Polynomial by Grouping

Once the GCF is factored out, we can often continue to _____ the polynomial, using a variety of techniques. We discuss here a technique for factoring polynomials called factoring by _____ .

To Factor a Four-Term Polynomial by Grouping

Step 1:

Step 2:

Step 3:

Step 4:

WORK WITH ME #1.

Factor each four-term polynomial by grouping.

a. $5xy - 15x - 6y + 18$

b. $6a^2 + 9ab^2 + 6ab + 9b^3$

Helpful Hint

One more reminder: When _____ a polynomial, make sure the polynomial is written as a product.

YOUR TURN #1:   **YOUR TURN #2:**

Section 13.2 Factor Trinomials of the Form $x^2 + bx + c$

Objectives
> A Factor Trinomials of the Form $x^2 + bx + c$
> B Factor Out the Greatest Common Factor and Then Factor a Trinomial of
>   the Form $x^2 + bx + c$

Directions: Complete your Interactive Organizer by filling in the blanks and solving exercises as you complete each screen of the Interactive Assignment.

- For **WORK WITH ME** exercises, follow along and write each step needed and shown to solve, including the final answer.
- For **YOUR TURN** exercises, write the exercise generated for you in MyLab Math, then "show your work" by writing each step needed to solve, including the final answer.

Objective A: Factor Trinomials of the Form $x^2 + bx + c$

Let's factor trinomials of the form $x^2 + bx + c$, such as

$$x^2 + 4x + 3, \quad x^2 - 8x + 15, \quad x^2 + 4x - 12, \quad r^2 - r - 42$$

Factoring means to write as a _____ . Also, factoring and multiplying are _____ processes.

Using the FOIL method of multiplying binomials, let's find the product: $(x + 3)(x + 1)$.

$$(x + 3)(x + 1) = x^2 + 1x + 3x + 3$$
$$=$$

WORK WITH ME #1.

Factor $x^2 + 4x + 3$.

Helpful Hint

Since multiplication is commutative, the factored form of $x^2 + 4x + 3$ can be written as either $(x + 1)(x + 3)$ or $(x + 3)(x + 1)$.

From the previous screens, how do we factor a trinomial of the form $x^2 + bx + c$?

Factoring a Trinomial of the Form $x^2 + bx + c$

The factored form of $x^2 + bx + c$ is

The product of these numbers is c.

$$x^2 + bx + c = (x + \square)(x + \square)$$

The sum of these numbers is b.

WORK WITH ME #2.

Factor each trinomial completely. If a polynomial cannot be factored, write "prime."

a. $x^2 + 7x + 6$ b. $x^2 - 8x + 15$

c. $x^2 - 3x - 18$ d. $x^2 - 3xy - 4y^2$

YOUR TURN #1:   **YOUR TURN #2:**

YOUR TURN #3:   **YOUR TURN #4:**

Objective B: Factor Out the Greatest Common Factor and Then Factor a Trinomial of the Form $x^2 + bx + c$

Watch the objective video.

***VIDEO* WORK WITH ME.**

$3x^2 + 9x - 30$ $\qquad\qquad\qquad\qquad 5x^3y - 25x^2y^2 - 120xy^3$

YOUR TURN #1: $\qquad\qquad\qquad\qquad$ **YOUR TURN #2:**

Section 13.3 Factoring Trinomials of the Form $ax^2 + bx + c$ and Perfect Square Trinomials

Objectives

> A   Factor Trinomials of the Form $ax^2 + bx + c$, Where $a \neq 1$
> B   Factor Out a GCF Before Factoring a Trinomial of the Form $ax^2 + bx + c$
> C   Factor Perfect Square Trinomials

Directions: Complete your Interactive Organizer by filling in the blanks and solving exercises as you complete each screen of the Interactive Assignment.

- For **WORK WITH ME** exercises, follow along and write each step needed and shown to solve, including the final answer.
- For **YOUR TURN** exercises, write the exercise generated for you in MyLab Math, then "show your work" by writing each step needed to solve, including the final answer.

Objective A: Factor Trinomials of the Form $ax^2 + bx + c$, Where $a \neq 1$

Watch the objective video.

To factor a polynomial, first check for a _____ .

VIDEO WORK WITH ME.

$10x^2 + 31x + 3$ $4x^2 - 8x - 21$

Order of factors makes no difference. (Multiplication is _____)

WORK WITH ME #1.

Factor $2x^2 + 13x - 7$

YOUR TURN #1:   YOUR TURN #2:

Objective B: Factor Out a GCF Before Factoring a Trinomial of the Form $ax^2 + bx + c$

WORK WITH ME #1.

Let's factor: $30x^3 + 38x^2 + 12x$

Helpful Hint

Don't forget to include the _____ factor in the factored form.

WORK WITH ME #2.

Factor each trinomial completely.

a. $4x^3 - 9x^2 - 9x$

b. $-14x^2 + 39x - 10$

YOUR TURN #1:   **YOUR TURN #2:**

Objective C: Factor Perfect Square Trinomials

A trinomial that is the square of a binomial is called a **perfect** _____ **trinomial.**

For example,

$(x + 3)^2 =$

Thus, _____ is a perfect square trinomial.

Factoring Perfect Square Trinomials

$a^2 + 2ab + b^2 =$
$a^2 - 2ab + b^2 =$

WORK WITH ME #1.

Factor $x^2 + 12x + 36$

WORK WITH ME #2.

Factor each perfect square trinomial completely.

a. $x^2 + 22x + 121$

b. $9x^2 - 24xy + 16y^2$

YOUR TURN #1:   **YOUR TURN #2:**

Section 13.4 Factoring Trinomials of the Form $ax^2 + bx + c$ by Grouping

Objectives
A Use the Grouping Method to Factor Trinomials of the Form $ax^2 + bx + c$

Directions: Complete your Interactive Organizer by filling in the blanks and solving exercises as you complete each screen of the Interactive Assignment.

- For **WORK WITH ME** exercises, follow along and write each step needed and shown to solve, including the final answer.
- For **YOUR TURN** exercises, write the exercise generated for you in MyLab Math, then "show your work" by writing each step needed to solve, including the final answer.

Objective A: Use the Grouping Method to Factor Trinomials of the Form $ax^2 + bx + c$

There is an alternative method that can be used to factor trinomials of the form $ax^2 + bx + c$, a 1. This method is called the _____ **method** because it uses factoring by _____.

To Factor Trinomials by Grouping
Step 1
Step 2
Step 3
Step 4

WORK WITH ME #1.

Factor each trinomial by grouping.

a. $21y^2 + 17y + 2$ b. $10x^2 - 9x + 2$ c. $12x^3 - 27x^2 - 27x$

YOUR TURN #1:   **YOUR TURN #2:**

Section 13.5 Factoring Binomials

Objectives
      A Factor the Difference of Two Squares
      B Factor the Sum or Difference of Two Cubes

Directions: Complete your Interactive Organizer by filling in the blanks and solving exercises as you complete each screen of the Interactive Assignment.

- For **WORK WITH ME** exercises, follow along and write each step needed and shown to solve, including the final answer.
- For **YOUR TURN** exercises, write the exercise generated for you in MyLab Math, then "show your work" by writing each step needed to solve, including the final answer.

Objective A: Factor the Difference of Two Squares

Let's multiply the sum and difference of the same two terms.

  For example, $x + 3$ and $x - 3$.

$$(x+3)(x-3) =$$

The binomial $x^2 - 9$ is called a difference of _____ .

If we _____ the pattern for this product of a sum and difference, we have a pattern (formula) to factor the binomial difference of squares.

Factoring the Difference of Two Squares
    $a^2 - b^2 =$

Helpful Hint

Since multiplication is _____ , remember that the order of factors does not matter.

In other words,

$$a^2 - b^2 = (a + b)(a - b) \text{ or } (a - b)(a + b)$$

WORK WITH ME #1.

Factor each binomial completely.
  a. $121m^2 - 100n^2$ b. $16r^2 + 1$

c. $xy^3 - 9xyz^2$ d. $49 - \dfrac{9}{25}m^2$

YOUR TURN #1:   **YOUR TURN #2:**

YOUR TURN #3:

Objective B: Factor the Sum or Difference of Two Cubes

Although the _____ of two squares usually does not factor, the sum or difference of two _____ can be factored.

Factoring the Sum or Difference of Two Cubes

$a^3 + b^3 =$

$a^3 - b^3 =$

Helpful Hint

Recall that "factor" means "to write as a _____."

These patterns for factoring sums and differences of cubes can be checked by _____.

WORK WITH ME #1.

Factor the sum of cubes: $x^3 + 125$

Helpful Hint

When factoring sums or difference of cubes, notice the sign patterns.

$$x^3 + y^3 = (x + y)(x^2 - xy + y^2)$$

same sign

opposite signs   always positive

$$x^3 - y^3 = (x - y)(x^2 + xy + y^2)$$

same sign

opposite signs   always positive

WORK WITH ME #2.

Factor.

a. $x^3 y^3 - 64$

b. $8m^3 + 64$

YOUR TURN #1:

YOUR TURN #2:

Section 13.6 Solving Quadratic Equations by Factoring

Objectives
- A Solve Quadratic Equations by Factoring
- B Solve Equations with Degree Greater than 2 by Factoring
- C Find the x-Intercepts of the Graph of a Quadratic Equation in Two Variables

---

Directions: Complete your Interactive Organizer by filling in the blanks and solving exercises as you complete each screen of the Interactive Assignment.
- For **WORK WITH ME** exercises, follow along and write each step needed and shown to solve, including the final answer.
- For **YOUR TURN** exercises, write the exercise generated for you in MyLab Math, then "show your work" by writing each step needed to solve, including the final answer.

Objective A: Solve Quadratic Equations by Factoring

Let's study a new type of equation—the _____ **equation**.

Quadratic Equation

A quadratic equation is one that can be written in the form

$$ax^2 + bx + c = 0$$

where a, b, and c are real numbers and $a \neq 0$.

The form $ax^2 + bx + c = 0$ is called the _____ **form** of a quadratic equation.

Show Me
  Which quadratic equations are in standard form?

$4x^2 - 28 = -49$ $x^2 - 9x - 22 = 0$ $x(2x - 7) = 4$

Some quadratic equations can be solved by making use of factoring and the _____ **factor property.**

Zero Factor Property

If a and b are real numbers and if $ab = 0$, then $a = 0$ or $b = 0$.

Helpful Hint

The zero factor property says that if a product is 0, then a _____ is 0.

If $a \cdot b = 0$, then $a = 0$ or $b = 0$.
If $x(x + 5) = 0$, then $x = 0$ or $x + 5 = 0$.
If $(x + 7)(2x - 3) = 0$, then $x + 7 = 0$ or $2x - 3 = 0$.

Use this property only when the product is 0.

WORK WITH ME #1.

Use the Zero Factor Property to solve $(x - 3)(x + 1) = 0$.

WORK WITH ME #2.

Solve each equation.

a. $x^2 + 2x - 8 = 0$

b. $(2x + 3)(4x - 5) = 0$

To Solve Quadratic Equations by Factoring

Step 1

Step 2

Step 3

Step 4

Step 5

WORK WITH ME #3.

Solve $x(3x - 1) = 14$.

YOUR TURN #1:　　　　　　　　　　　　　　　**YOUR TURN #2:**

YOUR TURN #3:

Objective B: Solve Equations with Degree Greater than 2 by Factoring

Watch the objective video.

VIDEO WORK WITH ME.

$(2x+3)(2x^2-5x-3)=0$

Zero Factor Theorem:

Factored Form that equals _____ .

YOUR TURN #1:

Objective C: Find the *x*-Intercepts of the Graph of a Quadratic Equation in Two Variables

Watch the objective video.

VIDEO WORK WITH ME.

$y=2x^2+11x-6$

To find *x*-intercepts let $y=$ _____ and solve for *x*.

YOUR TURN #1:

Section 13.7 Quadratic Equations and Problem Solving

Objectives
A Solve Problems That Can Be Modeled by Quadratic Equations

Directions: Complete your Interactive Organizer by filling in the blanks and solving exercises as you complete each screen of the Interactive Assignment.
- For **WORK WITH ME** exercises, follow along and write each step needed and shown to solve, including the final answer.
- For **YOUR TURN** exercises, write the exercise generated for you in MyLab Math, then "show your work" by writing each step needed to solve, including the final answer.

Objective A: Solve Problems That Can Be Modeled by Quadratic Equations

WORK WITH ME #1.

Solve.
  a. The *perimeter* of the triangle is 85 feet. Find the lengths of its sides.

  b. The product of two consecutive page numbers is 420. Find the page numbers.

WORK WITH ME #2.

Solve.
A piece of luggage is dropped from a cliff 256 feet above the ground. Neglecting air resistance, the height h in feet of the luggage above the ground after t seconds is given by the quadratic equation $h = -16t^2 + 256$. Find how long it takes for the luggage to hit the ground.

329

Recall that a _____ triangle is a triangle that contains a 90° or right angle. Let's review the **Pythagorean theorem**, which applies to right triangles only.

Pythagorean Theorem

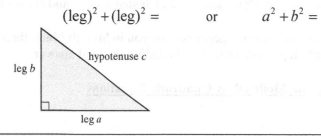

$$(\text{leg})^2 + (\text{leg})^2 = \qquad \text{or} \qquad a^2 + b^2 =$$

The **hypotenuse** of a right triangle is the side _____ the right angle and is the _____ side of the triangle.

The **legs** of a right triangle are the _____ sides of the triangle.

Helpful Hint

If you use this formula, don't forget that c represents the _____ of the hypotenuse.

WORK WITH ME #3.

Solve.
One leg of a right triangle is 4 millimeters longer than the smaller leg and the hypotenuse is 8 millimeters longer than the smaller leg. Find the lengths of the sides of the triangle.

YOUR TURN #1:   **YOUR TURN #2:**

YOUR TURN #3:

Chapter 13 Review and Practice

> Study Skills
> Chapter Vocabulary
> Getting Ready for the Test
> Review Exercises
> Practice Chapter Test

Study Skills

Directions: **Watch the Study Skills video.**

Chapter Vocabulary

WORK WITH ME.

Fill in each blank with one of the words or phrases listed below:

factoring	quadratic equation	perfect square trinomial	0
greatest common factor	hypotenuse	sum of two cubes	1
difference of two cubes	difference of two squares	triangle	leg

1. An equation that can be written in the form $ax^2 + bx + c = 0$ (with a not 0) is called a

_____ .

2. _____ is the process of writing an expression as a product.

3. The _____ of a list of terms is the product of all common factors.

4. A trinomial that is the square of some binomial is called a _____ .

5. The expression $a^2 - b^2$ is called a _____ .

6. The expression $a^3 - b^3$ is called a _____ .

7. The expression $a^3 + b^3$ is called a _____ .

8. By the zero factor property, if the product of two numbers is 0, then at least one of the numbers must be _____ .

9. In a right triangle, the side opposite the right angle is called the _____ .

10. In a right triangle, each side adjacent to the right angle is called a _____ .

11. The Pythagorean theorem states that $(\text{leg})^2 + (\text{leg})^2 = ($ _____ $)^2$.

Getting Ready for the Test.

- These exercises will help you avoid common errors while taking your chapter test.

General Directions: Read the exercise Write any notes or steps in this Interactive Organizer, along with your answer to the exercise. In the MyLab Math Interactive Assignment, click the **SHOW ANSWERS** button to check your answers. Correct any errors, or press the **PLAY** button for a video solution.

*All the exercises below are **Multiple Choice.** Choose the correct letter. Also, letters may be used more than once.*

1. The greatest common factor of the terms of $10x^4 - 70x^3 + 2x^2 - 14x$ is
   A. $2x^2$ B. $2x$ C. $7x^2$ D. $7x$

2. Choose the expression that is NOT a factored form of $9y^3 - 18y^2$.
   A. $9(y^3 - 2y^2)$ B. $9(y^2 - 2y)$ C. $9y^2(y - 2)$ D. $9 \cdot y^3 - 18 \cdot y^2$

Identify each expression as:
         A. A factored expression   *or*   B. Not a factored expression

3. $(x - 1)(x + 5)$

4. $z(z + 12)(z - 12)$

5. $y(x - 6) + 1(x - 6)$

6. $m \cdot m - 5 \cdot 5$

7. Choose the correct factored form for $4x^2 + 16$ or select "can't be factored."
   A. can't be factored   B. $4(x^2 + 4)$ C. $4(x + 2)^2$ D. $4(x + 2)(x - 2)$

8. Which of the binomials can't be factored using real numbers?
   A. $x^2 + 64$ B. $x^2 - 64$ C. $x^3 + 64$ D. $x^3 - 64$

9. To solve $x(x + 2) = 15$, which is an incorrect next step?
   A. $x^2 + 2x = 15$ B. $x(x + 2) - 15 = 0$ C. $x = 15$ and $x + 2 = 15$

Review Exercises

In the **MyLab Math, Interactive Assignment, Review Exercises** section, there are algorithmically generated "Your Turn" exercises so that you can check your knowledge of some core concepts in this chapter. Insert a few sheets of paper in your Interactive Organizer to "record and show your work" along with the final answer.

Practice Chapter Test

- These exercises will help you practice for your chapter test.

General Directions: For each exercise, "show your work" by writing each step in the solution process within your Interactive Organizer, including your final answer. In the MyLab Math Interactive Assignment, click the Show Answer button to check your answer. Correct any errors, or press the **PLAY** button for a video solution.

Factor each polynomial completely. If a polynomial cannot be factored, write "prime."

1. $x^2 + 11x + 28$

2. $49 - m^2$

3. $y^2 + 22y + 121$

4. $4(a + 3) - y(a + 3)$

5. $x^2 + 4$

6. $y^2 - 8y - 48$

7. $x^2 + x - 10$

8. $9x^3 + 39x^2 + 12x$

9. $3a^2 + 3ab - 7a - 7b$

10. $3x^2 - 5x + 2$

11. $x^2 + 14xy + 24y^2$

12. $180 - 5x^2$

13. $6t^2 - t - 5$

14. $xy^2 - 7y^2 - 4x + 28$

15. $x - x^5$

16. $-xy^3 - x^3y$

17. $64x^3 - 1$

18. $8y^3 - 64$

Solve each equation.

19. $(x - 3)(x + 9) = 0$

20. $x^2 + 5x = 14$

21. $x(x + 6) = 7$

22. $3x(2x - 3)(3x + 4) = 0$

23. $5t^3 - 45t = 0$

24. $t^2 - 2t - 15 = 0$

25. $6x^2 = 15x$

Solve each problem.

26. A deck for a home is in the shape of a triangle. The length of the base of the triangle is 9 feet longer than its altitude. If the area of the triangle is 68 square feet, find the length of the base.

27. The sum of two numbers is 17 and the sum of their squares is 145. Find the numbers.

28. An object is dropped from the top of the Woolworth Building on Broadway in New York City. The height h of the object after t seconds is given by the equation

$$h = -16t^2 + 784$$

Find how many seconds pass before the object reaches the ground.

29. Find the lengths of the sides of a right triangle if the hypotenuse is 10 centimeters longer than the shorter leg and 5 centimeters longer than the longer leg.

Section 14.1 Rational Functions and Simplifying Rational Expressions

Objectives
- A   Find the Domain of a Rational Function
- B   Simplify or Write Rational Expressions in Lowest Terms
- C   Write Equivalent Rational Expressions of the Form $-\dfrac{a}{b} = \dfrac{-a}{b} = \dfrac{a}{-b}$
- D   Use Rational Functions in Applications

Directions: Complete your Interactive Organizer by filling in the blanks and solving exercises as you complete each screen of the Interactive Assignment.

- For **WORK WITH ME** exercises, follow along and write each step needed and shown to solve, including the final answer.
- For **YOUR TURN** exercises, write the exercise generated for you in MyLab Math, then "show your work" by writing each step needed to solve, including the final answer.

Objective A: Find the Domain of a Rational Function

Watch the objective video.

A _____ expression can be written as $\dfrac{P}{Q}$, where P and Q are polynomials, $Q \neq 0$.

A rational _____ is of the form $f(x) = \dfrac{P}{Q}$ as long as $\dfrac{P}{Q}$ is a rational expression.

***VIDEO* WORK WITH ME.**

$$f(x) = \frac{3x}{7-x}$$

$$C(x) = \frac{x+3}{x^2 - 4}$$

YOUR TURN #1:   **YOUR TURN #2:**

Objective B: Simplify or Write Rational Expressions in Lowest Terms

To _____ a rational expression, or to write it in _____ terms, we use a method similar to simplifying a fraction – we "_____ factors of 1."

Simplify $\dfrac{15}{65}$:

$$\frac{15}{65} = \frac{3 \cdot 5}{13 \cdot 5} = \frac{3}{13} \cdot \frac{5}{5} = \frac{3}{13} \cdot 1 = \frac{3}{13}$$

This is called "removing a factor of 1."

Let's simplify the rational expression $\dfrac{x^2 - 9}{x^2 + x - 6}$.

$$\frac{x^2 - 9}{x^2 + x - 6}$$

$$= \frac{(x-3)(x+3)}{(x-2)(x+3)} \quad \text{Factor the } \underline{\hspace{2cm}} \text{ and } \underline{\hspace{2cm}}.$$

$$= \frac{(x-3)(x+3)}{(x-2)(x+3)} \quad \text{Look for } \underline{\hspace{1.5cm}} \text{ factors.}$$

$$= \frac{x-3}{x-2} \cdot \frac{x+3}{x+3} \quad \text{Separate the common } \underline{\hspace{1.5cm}}.$$

$$= \frac{x-3}{x-2} \cdot 1 \quad \text{Write } \frac{x+3}{x+3} \text{ as } \underline{\hspace{0.8cm}}.$$

$$= \frac{x-3}{x-2} \quad \text{Multiply to } \underline{\hspace{2cm}} \text{ a factor of 1.}$$

Fundamental Principle of Rational Expressions

For any rational expression $\dfrac{P}{Q}$ and any polynomial R, where $R \neq 0$,

$$\frac{PR}{QR} = \frac{P}{Q} \cdot \frac{R}{R} = \frac{P}{Q} \cdot 1 = \frac{P}{Q}$$

or, simply,

$$\frac{PR}{QR} = \frac{P}{Q}$$

The following steps may be used to simplify rational expressions or to write a rational expression in lowest terms.

Simplifying or Writing a Rational Expression in Lowest Terms

Step 1:

Step 2:

WORK WITH ME #1.

Simplify each expression.

a. $\dfrac{-5a-5b}{a+b}$

b. $\dfrac{x+7}{7+x}$

c. $\dfrac{x-7}{7-x}$

d. $\dfrac{4-x^2}{x-2}$

YOUR TURN #1:   **YOUR TURN #2:**

YOUR TURN #3:

Objective C: Write Equivalent Rational Expressions of the Form $-\dfrac{a}{b} = \dfrac{-a}{b} = \dfrac{a}{-b}$

Watch the objective video.

VIDEO WORK WITH ME.

$-\dfrac{x+11}{x-4}$

YOUR TURN #1:

Objective D: Use Rational Functions in Applications

Watch the objective video.

VIDEO WORK WITH ME.

The total revenue from the sale of a popular book is approximated by the rational function

$$R(x) = \dfrac{1000x^2}{x^2 + 4}$$

where x is the number of years since publication and $R(x)$ is the total revenue in millions of dollars.

a. Find the total revenue at the end of the first year.
b. Find the total revenue at the end of the second year.
c. Find the revenue during the second year only.
d. Find the domain of the function R.

YOUR TURN #1:

Section 14.2 Multiplying and Dividing Rational Expressions

Objectives
   A Multiply Rational Expressions
   B Divide Rational Expressions
   C Multiply or Divide Rational Expressions
   D Convert Between Units of Measure

Directions: Complete your Interactive Organizer by filling in the blanks and solving exercises as you complete each screen of the Interactive Assignment.
   - For **WORK WITH ME** exercises, follow along and write each step needed and shown to solve, including the final answer.
   - For **YOUR TURN** exercises, write the exercise generated for you in MyLab Math, then "show your work" by writing each step needed to solve, including the final answer.

Objective A: Multiply Rational Expressions

WORK WITH ME #1.

Multiply: $\dfrac{3}{5} \cdot \dfrac{10}{11}$

Multiply: $\dfrac{x-3}{x+5} \cdot \dfrac{2x+10}{x^2-9}$

Multiplying Rational Expressions

If $\dfrac{P}{Q}$ and $\dfrac{R}{S}$ are rational expressions, then

$$\dfrac{P}{Q} \cdot \dfrac{R}{S} =$$

To multiply rational expressions, multiply the _____ and multiply the _____.

Multiplying Rational Expressions

Step 1:

Step 2:

Step 3:

WORK WITH ME #2.

Multiply.

$$\frac{5x-20}{3x^2+x} \cdot \frac{3x^2+13x+4}{x^2-16}$$

YOUR TURN #1:

Objective B: Divide Rational Expressions

Watch the objective video.

Dividing Rational Expressions

If $\dfrac{P}{Q}$ and $\dfrac{R}{S}$ are rational expressions and $\dfrac{R}{S}$ is not 0, then

$$\frac{P}{Q} \div \frac{R}{S} =$$

To Divide: _____ by its _____.

***VIDEO* WORK WITH ME.**

$$\frac{x+2}{7-x} \div \frac{x^2-5x+6}{x^2-9x+14}$$

YOUR TURN #1:

Objective C: Multiply or Divide Rational Expressions

Watch the objective video.

VIDEO WORK WITH ME.

$$\frac{5x-10}{12} \div \frac{4x-8}{8}$$

$$\frac{2x}{x^2-25} \text{ ft}$$

$$\frac{x+5}{9x} \text{ ft}$$

YOUR TURN #1:   **YOUR TURN #2:**

Objective D: Convert Between Units of Measure

Watch the objective video.

VIDEO WORK WITH ME.

3 cubic yards = _____ cubic feet

YOUR TURN #1:

Section 14.3 Adding and Subtracting Rational Expressions with Common Denominators and Least Common Denominator

Objectives

    A Add and Subtract Rational Expressions with the Same Denominator
    B Find the Least Common Denominator of a List of Rational Expressions
    C Write a Rational Expression as an Equivalent Expression Whose Denominator is Given

Directions: Complete your Interactive Organizer by filling in the blanks and solving exercises as you complete each screen of the Interactive Assignment.

- For **WORK WITH ME** exercises, follow along and write each step needed and shown to solve, including the final answer.
- For **YOUR TURN** exercises, write the exercise generated for you in MyLab Math, then "show your work" by writing each step needed to solve, including the final answer.

Objective A: Add and Subtract Rational Expressions with the Same Denominator

WORK WITH ME #1.

Add: $\dfrac{6}{5} + \dfrac{2}{5}$

Add: $\dfrac{9}{x+2} + \dfrac{3}{x+2}$

Adding and Subtracting Rational Expressions with Common Denominators

If $\dfrac{P}{R}$ and $\dfrac{Q}{R}$ are rational expressions, then

$$\frac{P}{R} + \frac{Q}{R} = \qquad \text{and} \qquad \frac{P}{R} - \frac{Q}{R} =$$

To add or subtract rational expressions, add or subtract the _____ and place the sum or difference over the _____ denominator.

WORK WITH ME #2.

Add or subtract as indicated. Simplify the result if possible.

a. $\dfrac{9}{3+y} + \dfrac{y+1}{3+y}$

b. $\dfrac{2x+3}{x^2 - x - 30} - \dfrac{x-2}{x^2 - x - 30}$

YOUR TURN #1:   **YOUR TURN #2:**

Objective B: Find the Least Common Denominator of a List of Rational Expressions

To add or subtract rational expressions with _____ denominators, we first find the least common denominator (LCD) and then write all rational expressions as equivalent expressions with the _____ .

Finding the Least Common Denominator (LCD)

Step 1:

Step 2:

WORK WITH ME #1.

Find the LCD of $\dfrac{6m^2}{3m+15}$, $\dfrac{2}{(m+5)^2}$

WORK WITH ME #2.

Find the LCD for the list of rational expressions.

$\dfrac{9}{8x}$, $\dfrac{3}{2x+4}$

YOUR TURN #1:   **YOUR TURN #2:**

Objective C: Write a Rational Expression as an Equivalent Expression Whose Denominator is Given

Watch the objective video.

VIDEO WORK WITH ME.

$$\frac{6}{3a} = \frac{}{12ab^2}$$
$$\frac{9a+2}{5a+10} = \frac{}{5b(a+2)}$$

YOUR TURN #1:   **YOUR TURN #2:**

Section 14.4 Adding and Subtracting Rational Expressions with Unlike Denominators

Objectives
A Add and Subtract Rational Expressions with Unlike Denominators

Directions: Complete your Interactive Organizer by filling in the blanks and solving exercises as you complete each screen of the Interactive Assignment.

- For **WORK WITH ME** exercises, follow along and write each step needed and shown to solve, including the final answer.
- For **YOUR TURN** exercises, write the exercise generated for you in MyLab Math, then "show your work" by writing each step needed to solve, including the final answer.

Objective A: Add and Subtract Rational Expressions with Unlike Denominators

Adding or Subtracting Rational Expression with Unlike Denominators
Step 1:
Step 2:
Step 3:
Step 4:

WORK WITH ME #1.

$$\frac{5x+7}{x} - \frac{x^2+7x}{x^2}$$

WORK WITH ME #2.

a. $\dfrac{3a}{2a+6} - \dfrac{a-1}{a+3}$

b. $\dfrac{x+8}{x^2-5x-6} + \dfrac{x+1}{x^2-4x-5}$

YOUR TURN #1:   **YOUR TURN #2:**

YOUR TURN #3:   **YOUR TURN #4:**

Section 14.5 Solving Equations Containing Rational Expressions

Objectives
    A Solve Equations Containing Rational Expressions
    B Solve Equations Containing Rational Expressions for a Specified Variable

Directions: Complete your Interactive Organizer by filling in the blanks and solving exercises as you complete each screen of the Interactive Assignment.
- For **WORK WITH ME** exercises, follow along and write each step needed and shown to solve, including the final answer.
- For **YOUR TURN** exercises, write the exercise generated for you in MyLab Math, then "show your work" by writing each step needed to solve, including the final answer.

Objective A: Solve Equations Containing Rational Expressions

Examples of Equations Containing Rational Expressions

$$\frac{x}{2} + \frac{8}{3} = \frac{1}{6} \text{ and } \frac{4x}{x^2 + x - 30} + \frac{2}{x - 5} = \frac{1}{x + 6}$$

To solve equations such as these, use the multiplication property of equality to clear the equation of fractions by multiplying both sides of the equation by the LCD.

WORK WITH ME #1.

Solve the equation.
$$\frac{2}{y} + \frac{1}{2} = \frac{5}{2y}$$

WORK WITH ME #2.

Solve the equation.
$$3 - \frac{6}{x} = x + 8$$

Solving an Equation Containing Rational Expressions

Step 1:

Step 2:

Step 3:

WORK WITH ME #3.

Solve each equation.

a. $2 + \dfrac{3}{a-3} = \dfrac{a}{a-3}$

b. $\dfrac{4r-4}{r^2+5r-14} + \dfrac{2}{r+7} = \dfrac{1}{r-2}$

YOUR TURN #1:   **YOUR TURN #2:**

YOUR TURN #3:

Objective B: Solve Equations Containing Rational Expressions for a Specified Variable

Watch the objective video.

VIDEO WORK WITH ME.

$T = \dfrac{2u}{B+E}$

YOUR TURN #1:   **YOUR TURN #2:**

Section 14.6 Problem Solving with Proportions and Rational Expressions

Objectives
A Use Proportions to Solve Problems
B Solve Problems About Numbers
C Solve Problems About Work
D Solve Problems About Distance

Directions: Complete your Interactive Organizer by filling in the blanks and solving exercises as you complete each screen of the Interactive Assignment.

- For **WORK WITH ME** exercises, follow along and write each step needed and shown to solve, including the final answer.
- For **YOUR TURN** exercises, write the exercise generated for you in MyLab Math, then "show your work" by writing each step needed to solve, including the final answer.

Objective A: Use Proportions to Solve Problems

Watch the objective video.

A _____ is a statement that two ratios are _____ .

Cross Products
If $\dfrac{a}{b} = \dfrac{c}{d}$ then $ad =$

VIDEO WORK WITH ME.

There are 110 calories per 28.8 grams of Frosted Flakes cereal. Find how many calories are in 43.2 grams of this cereal.

VIDEO WORK WITH ME.

Here are two similar triangles.

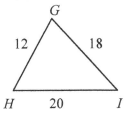
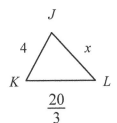

YOUR TURN #1:   **YOUR TURN #2:**

Objective B: Solve Problems About Numbers

Watch the objective video.

VIDEO WORK WITH ME.

Twelve divided by the sum of x and 2 equals the quotient of 4 and the difference of x and 2. Find x.

YOUR TURN #1:

Objective C: Solve Problems About Work

Watch the objective video.

VIDEO WORK WITH ME.

In 2 minutes, a conveyor belt moves 300 pounds of recyclable aluminum from the delivery truck to a storage area. A smaller belt moves the same quantity of cans the same distance in 6 minutes. If both belts are used, find how long it takes to move the cans to the storage area.

YOUR TURN #1:

Objective D: Solve Problems About Distance

Watch the objective video.

VIDEO WORK WITH ME.

A car travels 280 miles in the same time that a motorcycle travels 240 miles. If the car's speed is 10 miles per hour more than the motorcycle's, find the speed of the car and the speed of the motorcycle.

YOUR TURN #1:

Section 14.7 Simplifying Complex Fractions

Objectives

    A Simplify Complex Fractions by Simplifying the Numerator and Denominator and Then Dividing

    B Simplify Complex Fractions by Multiplying by a Common Denominator

    C Simplify Expressions with Negative Exponents

Directions: Complete your Interactive Organizer by filling in the blanks and solving exercises as you complete each screen of the Interactive Assignment.

- For **WORK WITH ME** exercises, follow along and write each step needed and shown to solve, including the final answer.
- For **YOUR TURN** exercises, write the exercise generated for you in MyLab Math, then "show your work" by writing each step needed to solve, including the final answer.

Objective A: Simplify Complex Fractions by Simplifying the Numerator and Denominator and Then Dividing

A rational expression whose numerator, denominator, or both contain one or more rational expressions is called a _____ rational expression or a _____ fraction.

Examples of Complex Fractions

$$\frac{\dfrac{1}{a}}{\dfrac{b}{2}} \qquad \frac{\dfrac{x}{2y^2}}{\dfrac{6x-2}{9y}} \qquad \frac{x+\dfrac{1}{y}}{y+1}$$

Our goal is to _____ complex fractions. A complex fraction is simplified when it is in the form $\dfrac{P}{Q}$, where P and Q are polynomials that have _____ common factors.

Simplifying a Complex Fraction: Method 1

Step 1:

Step 2:

Step 3:

WORK WITH ME #1.

Simplify the complex fraction.

$$\frac{\frac{10}{3x}}{\frac{5}{6x}}$$

WORK WITH ME #2.

Simplify the complex fraction.

$$\frac{\frac{1}{x} - \frac{1}{2}}{\frac{1}{3} - \frac{x}{6}}$$

YOUR TURN #1:

Objective B: Simplify Complex Fractions by Multiplying by a Common Denominator

Now we look at another _____ of simplifying complex fractions. With this method, we multiply the numerator and the denominator of the complex fraction by the _____ of all fractions in the complex fraction.

Simplifying a Complex Fraction: Method 2

Step 1:

Step 2:

WORK WITH ME #1.

Simplify the complex fraction.

$$\dfrac{\dfrac{x+2}{x} - \dfrac{2}{x-1}}{\dfrac{x+1}{x} + \dfrac{x+1}{x-1}}$$

WORK WITH ME #2.

Simplify the complex fraction.

$$\dfrac{\dfrac{x}{y} + \dfrac{3}{2x}}{\dfrac{x}{2} + y}$$

YOUR TURN #1:

Objective C: Simplify Expressions with Negative Exponents

Watch the objective video.

***VIDEO* WORK WITH ME.**

$$\dfrac{2a^{-1} + 3b^{-2}}{a^{-1} - b^{-1}}$$

YOUR TURN #1:

Chapter 14 Review and Practice

> Study Skills
> Chapter Vocabulary
> Getting Ready for the Test
> Review Exercises
> Practice Chapter Test

Study Skills

Directions: **Watch the Study Skills video.**

Chapter Vocabulary

WORK WITH ME.

Fill in each blank with one of the words or phrases listed below:

| least common denominator | simplifying | reciprocals | numerator | $\dfrac{-a}{b}$ | $\dfrac{a}{-b}$ |

| cross products | ratio | proportion |

| rational expression | domain | complex fraction | $\dfrac{-a}{-b}$ |

1. A(n) _____ is an expression that can be written in the form $\dfrac{P}{Q}$, where P and Q are polynomials and Q is not 0.

2. In a(n) _____ , the numerator and denominator or both may contain fractions.

3. For a rational expression $-\dfrac{a}{b} = $ _____ $ = $ _____ .

4. A rational expression is undefined when the _____ is 0.

5. The process of writing a rational expression in lowest terms is called _____ .

6. The expressions $\dfrac{2x}{7}$ and $\dfrac{7}{2x}$ are called _____ .

7. The _____ of a list of rational expressions is a polynomial of least degree whose factors include all factors of the denominators in the list.

8. A(n) _____ is the quotient of two numbers.

9. $\dfrac{x}{2} = \dfrac{7}{16}$ is an example of a(n) _____ .

10. If $\dfrac{a}{b} = \dfrac{c}{d}$, then *ad* and *bc* are called_____ .

11. The _____ of the rational function $f(x) = \dfrac{1}{x-3}$ is $\{x \mid x \text{ is a real number}, x \ne 3\}$.

Getting Ready for the Test.

- These exercises will help you avoid common errors while taking your chapter test.

General Directions: Read the exercise Write any notes or steps in this Interactive Organizer, along with your answer to the exercise. In the MyLab Math Interactive Assignment, click the **SHOW ANSWERS** button to check your answers. Correct any errors, or press the **PLAY** button for a video solution.

Multiple Choice. *Select the correct choice.*

1. Choose the expression that is equivalent to $\dfrac{-x}{4-x}$.

   A. $\dfrac{1}{4}$ B. $-\dfrac{1}{4}$ C. $\dfrac{x}{4-x}$ D. $\dfrac{x}{x-4}$

2. For which of these values (if any) is the expression $\dfrac{x+3}{x^2+9}$ undefined?

   A. –3   B. 3   C. –3 and 3   D. 0   E. none of these

Matching. *Match each rational expression to its simplified form. Letters may be used more than once.*

3. $\dfrac{y-6}{6-y}$ 4. $\dfrac{y+3}{3+y}$ A. 1

5. $\dfrac{x-2}{-2+x}$ 6. $\dfrac{m-4}{m+4}$ B. –1

                                                      C. neither 1 nor –1

Multiple Choice. *Select the correct choice.*

7. $\dfrac{8}{x^2} \cdot \dfrac{4}{x^2} =$

     A. $\dfrac{32}{x^2}$ B. $\dfrac{2}{x^2}$ C. $\dfrac{32}{x^4}$ D. 2   E. $\dfrac{1}{2}$

8. $\dfrac{8}{x^2} \div \dfrac{4}{x^2} =$

     A. $\dfrac{32}{x^2}$ B. $\dfrac{2}{x^2}$ C. $\dfrac{32}{x^4}$ D. 2   E. $\dfrac{1}{2}$

9. $\dfrac{8}{x^2} + \dfrac{4}{x^2} =$

     A. $\dfrac{32}{x^2}$ B. $\dfrac{2}{x^2}$ C. $\dfrac{12}{x^4}$ D. $\dfrac{12}{x^2}$

10. $\dfrac{7x}{x-1} - \dfrac{5+2x}{x-1} =$

     A. 5   B. $\dfrac{9x-5}{x-1}$ C. $\dfrac{5}{x-1}$ D. $\dfrac{14}{x-1}$

11. The LCD of $\dfrac{9}{25x}$ and $\dfrac{z}{10x^3}$ is

     A. $250x^4$ B. $250x$ C. $50x^4$ D. $50x^3$

12. The LCD of $\dfrac{5}{4x+8}$ and $\dfrac{9}{8x-8}$ is

     A. $(4x+8)(8x-8)$ B. $32(x+2)(x-1)$ C. $4(x+2)(x-1)$ D. $8(x+2)(x-1)$

Multiple Choice. *Identify each as an* **A.** expression *or* **B.** equation. *Letters may be used more than once or not at all.*

13. $\dfrac{5}{x} + \dfrac{1}{3}$ 14. $\dfrac{5}{x} + \dfrac{1}{3} = \dfrac{2}{x}$ 15. $\dfrac{a+5}{11} = 9$ 16. $\dfrac{a+5}{11} \cdot 9$

Multiple Choice. *Select the correct choice.*

17. Multiple the given equation through by the LCD of its terms. Choose the correct equivalent once this is done. Given equation: $\dfrac{x+3}{4} + \dfrac{5}{6} = 3$

     A. $(x+3)+5=3$ B. $3(x+3)+2\cdot 5=3$ C. $3(x+3)+2\cdot 5=12\cdot 3$ D. $6(x+3)+4\cdot 5=3$

357

18. Multiple the given equation through by the LCD of its terms. Choose the correct equivalent once this is done. Given equation: $3 - \dfrac{10x}{4(x+1)} = \dfrac{5}{6(x+1)}$

A. $3 - 10x = 5$ B. $3 - 3 \cdot 10x = 2 \cdot 5$ C. $3 \cdot 12(x+1) - 3 \cdot 10x = 2 \cdot 5$ D. $4(x+1) - 3 \cdot 10x = 2 \cdot 5$

19. Translate into an equation. Let x be the unknown number. "The quotient of a number and 5 equals the sum of that number and 12."

A. $\dfrac{x}{5} = x + 12$ B. $\dfrac{5}{x} = x + 12$ C. $\dfrac{x}{5} = x \cdot 12$ D. $\dfrac{x}{5} \cdot (x + 12)$

20. Write $\dfrac{2x^{-1}}{y^{-2} + (5x)^{-1}}$ without negative exponents.

A. $\dfrac{\dfrac{2}{x}}{\dfrac{1}{y^2} + \dfrac{1}{5x}}$ B. $\dfrac{\dfrac{1}{2x}}{\dfrac{1}{y^2} + \dfrac{1}{5x}}$ C. $\dfrac{y^2 + 5x}{2x}$ D. $\dfrac{\dfrac{2}{x}}{\dfrac{1}{y^2} + \dfrac{5}{x}}$

Review Exercises

In the **MyLab Math, Interactive Assignment, Review Exercises** section, there are algorithmically generated "Your Turn" exercises so that you can check your knowledge of some core concepts in this chapter. Insert a few sheets of paper in your Interactive Organizer to "record and show your work" along with the final answer.

Practice Chapter Test

- These exercises will help you practice for your chapter test.

General Directions: For each exercise, "show your work" by writing each step in the solution process within your Interactive Organizer, including your final answer. In the MyLab Math Interactive Assignment, click the Show Answer button to check your answer. Correct any errors, or press the **PLAY** button for a video solution.

1. Find the domain of the rational function.
$$g(x) = \frac{9x^2 - 9}{x^2 + 4x + 3}$$

2. For a certain computer desk, the average cost C (in dollars) per desk manufactured is
$$C(x) = \frac{100x + 3000}{x}$$
where x is the number of desks manufactured.
   a. Find the average cost per desk when manufacturing 200 computer desks.
   b. Find the average cost per desk when manufacturing 1000 computer desks.

Simplify each rational expression.

3. $\dfrac{3x-6}{5x-10}$

4. $\dfrac{x+6}{x^2+12x+36}$

5. $\dfrac{x+3}{x^3+27}$

6. $\dfrac{2m^3-2m^2-12m}{m^2-5m+6}$

7. $\dfrac{ay+3a+2y+6}{ay+3a+5y+15}$

8. $\dfrac{y-x}{x^2-y^2}$

Perform the indicated operation and simplify if possible.

9. $\dfrac{3}{x-1}\cdot(5x-5)$

10. $\dfrac{y^2-5y+6}{2y+4}\cdot\dfrac{y+2}{2y-6}$

11. $\dfrac{15x}{2x+5}-\dfrac{6-4x}{2x+5}$

12. $\dfrac{5a}{a^2-a-6}-\dfrac{2}{a-3}$

13. $\dfrac{6}{x^2-1}+\dfrac{3}{x+1}$

14. $\dfrac{x^2-9}{x^2-3x}\div\dfrac{xy+5x+3y+15}{2x+10}$

15. $\dfrac{x+2}{x^2+11x+18}+\dfrac{5}{x^2-3x-10}$

Solve each equation.

16. $\dfrac{4}{y}-\dfrac{5}{3}=-\dfrac{1}{5}$

17. $\dfrac{5}{y+1}=\dfrac{4}{y+2}$

18. $\dfrac{a}{a-3}=\dfrac{3}{a-3}-\dfrac{3}{2}$

19. $x-\dfrac{14}{x-1}=4-\dfrac{2x}{x-1}$

20. $\dfrac{10}{x^2-25}=\dfrac{3}{x+5}+\dfrac{1}{x-5}$

Simplify each complex fraction.

21. $\dfrac{\dfrac{5x^2}{yz^2}}{\dfrac{10x}{z^3}}$

22. $\dfrac{5-\dfrac{1}{y^2}}{\dfrac{1}{y}+\dfrac{2}{y^2}}$

23. $\dfrac{\dfrac{b}{a}-\dfrac{a}{b}}{\dfrac{1}{b}+\dfrac{1}{a}}$

24. In a sample of 85 fluorescent bulbs, 3 were found to be defective. At this rate, how many defective bulbs should be found in 510 bulbs?

25. One number plus five times its reciprocal is equal to six. Find the number.

26. A pleasure boat traveling down the Red River takes the same time to go 14 miles upstream as it takes to go 16 miles downstream. If the current of the river is 2 miles per hour, find the speed of the boat in still water.

27. An inlet pipe can fill a tank in 12 hours. A second pipe can fill the tank in 15 hours. If both pipes are used, find how long it takes to fill the tank.

28. Given that the two triangles are similar, find x.

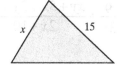